# Business Event Legacies:
## Global industry case studies

Carmel Foley, Deborah Edwards,
Karin Weber

(G) Goodfellow Publishers Ltd

(G) Published by Goodfellow Publishers Limited,
26 Home Close, Wolvercote, Oxford OX2 8PS
http://www.goodfellowpublishers.com

British Library Cataloguing in Publication Data: a catalogue record for this title is available from the British Library.

Library of Congress Catalog Card Number: on file.

ISBN: 978-1-911635-29-1

DOI: 110.23912/9781911635291-4966



Design and typesetting by P.K. McBride, www.macbride.org.uk

Cover design by Cylinder

Cover image credit: https://www.istockphoto.com/portfolio/Giii

Printed by Lightning Source

# Contents


| | | |
|---|---|---|
| | Foreword | v |
| | About the authors | vii |
| **1** | **Introduction** | **1** |
| | A time to reflect | 2 |
| | The tourism/events partnership | 3 |
| | Looking ahead | 5 |
| | In summary | 12 |
| **2** | **Conferences as Catalysts for Thriving Economies** | **16** |
| | Case study: The value of conferences beyond tourism benefits | 17 |
| | Theoretical discussion | 31 |
| | Lessons learned | 41 |
| **3** | **Collaboration, Partnerships and Sustainable Development** | **46** |
| | Case study: Business event legacies for developing nations – ICCA Congress, Malaysia | 47 |
| | Theoretical discussion | 58 |
| | Lessons learned | 71 |
| **4** | **The Economic Legacies of Business Events – Measuring the Iceberg** | **75** |
| | Case study: Economic legacies of London Tech Week | 77 |
| | Theoretical discussion | 89 |
| | Lessons learned | 100 |
| **5** | **Organisational Development Depends on Knowledge Sharing** | **104** |
| | Case study: Knowledge sharing and organisational development through the conference EuroHeartCare 2017 | 105 |
| | Theoretical discussion | 118 |
| | Lessons learned | 126 |
| **6** | **Legacies through Thoughtful Supply Chains** | **132** |
| | Case study: ICC Sydney: Feeding Your Performance | 133 |
| | Theoretical discussion | 152 |
| | Lessons learned | 156 |

**7 Academic Scholars and Conferences** **165**

Case study: Benefits and challenges of hosting conferences for academics and universities 167

Theoretical discussion 185

Lessons learned 193

**8 Business Events and Innovation** **197**

Case study: Swiss Fintech Start-up Corner 199

Theoretical discussion 213

Lessons learned 216

**9 Business Events and Diplomacy** **221**

Case study: Investing in business events for regional growth and development 222

Theoretical discussion 234

Lessons learned 239

**10 Leveraging Legacy** **242**

How legacies were leveraged in the case studies 243

Challenges faced in leveraging legacies 246

A way forward: The seven-step framework for business event legacy planning 252

Leveraging legacy 254

Conclusion 260

**Index** **263**

# Foreword

A defining issue for the meetings industry over the past decade has been a fundamental transition of its value proposition from one based essentially on travel-related benefits to one that instead recognizes its real value as being associated with the economic, professional, academic and social benefits arising from these events. This was hardly news to organizers but required a big shift in thinking for many host destinations and suppliers.

The problem was – and remains – that the capability for measuring these more diffuse benefits has not kept up with changing attitudes. Most jurisdictions still rely on what are essentially visitor economy measures, such as direct revenues, hotel room-nights and economic impact measures, for the simple reason that these are easy to calculate and report on, in terms of justifying what for most destination is a major investment in the infrastructure and ongoing promotion required to compete. However, the result is an ongoing sense – particularly amongst those in the tourism sector – that these are the only 'real' values associated with this industry.

Even though more governments – and certainly organizers – now accept that the greatest legacies of meetings, conventions and exhibitions are derived from the kinds of exchanges and interactions that take place there, our collective grip on corresponding measures still has a long way to go, for some understandable reasons. For one, many legacies are not generated immediately, and instead only reveal themselves completely over time. For another, as critical as advances in economic prospects, academic advancement and professional capabilities may be, they are essentially qualitative measures and certainly don't lend themselves to the kind of monetization that would facilitate direct comparison with spending or economic impact numbers.

Enter the incredibly valuable work undertaken by the University of Technology Sydney. Working closely over many years with pioneering organizations like Business Events Sydney and the Joint Meetings Industry Council, Carmel Foley and Deborah Edwards have developed and advanced measurement of the values associated with business events in ways that recognize their often-elusive but nevertheless important values. At the same time, they have helped to create frameworks in which these values can be better demonstrated and documented. In doing so, they have benefited not only an industry that needs such measures to support their ongoing role in their respective communities but assisted event organizers and their related associations and industries, many of whom have been struggling with the same issues.

It is an ongoing, but hugely important process. The fact that recognizable values are difficult to quantify with precision should not be an excuse to ignore them, and yet that is exactly what has happened for many years. By accepting the importance of clarifying and illustrating the broader benefits, as well as a long-term view of how those benefits are often only achieved incrementally through the kinds of interactions and exchanges that only take place in the collegial environment of business events, Edwards and Foley (UTS) have been major players in facilitating a necessary transition.

The contents of this publication illustrate the progress that has been made and, equally important, the further steps that are necessary to achieve the legitimacy these new event measures need and deserve. From an industry perspective, we highly recommend *Business Event Legacies* for the insights it provides into what will likely remain the top priority for the business events sector for some time to come.

Rod Cameron, Executive Director,

Joint Meetings Industry Council

# About the authors

**Carmel Foley** is Associate Professor in Event Management with the UTS Business School at the University of Technology Sydney. She is internationally recognised for her business events research which has influenced public policy in Australia and redefined how business events are valued and measured globally. Key industry projects delivered with her co-investigator Associate Professor Deborah Edwards include the *Beyond Tourism Benefits* research series commissioned by Business Events Sydney and the *Value of Meetings* case study project commissioned by the Joint Meetings Industry Council. Carmel applies her expertise in developing economic and social impact measurement methodologies in a range of event, tourism, sport and leisure contexts and her work is underpinned by a mission to support the economic, social and cultural prosperity of our communities.

**Deborah Edwards** is an Associate Professor and Adjunct Fellow with the Business School, University of Technology Sydney. Deborah has varied research interests including business events, tourism sustainability, and visitor experiences. She has a significant number of publications with her co-author Associate Professor Carmel Foley with whom she wrote *The Power of Conferences, Stories of serendipity, Innovation and Driving Social Change*. This book explores the long-term impacts of conferences through the stories of 10 prominent Australians who are synonymous with important societal outcomes. *The Power of Conferences* explains how unexpected encounters at conferences have led to significant global discoveries.

**Karin Weber** is an Associate Professor in the School of Hotel & Tourism Management at the Hong Kong Polytechnic University. She has published widely in her two major areas of research interest – service failure/recovery and business events; her research has been recognized with numerous international awards. Karin has served in various editorial capacities (co-editor, associate editor, coordinating editor, regional editor) for event-related journals in addition to serving on editorial boards of top tourism/hospitality journals. She was the lead editor and chapter author of an early book on convention tourism published by Haworth Press New York in 2002, and chaired the first Convention and Expo Summit in Asia in 2003. She also consulted for various international business event industry associations. Karin was a member of the international Advisory Panel for the Joint Meeting Industry Council's *Value of Meetings* case study project, led by Carmel Foley and Deborah Edwards.

# 1 Introduction

> *Whether to accelerate scientific research, raise funding, or drive social change, the stories and individuals featured have been captured to inspire others to take the long view and think about how they can use their events to help create a lasting impact in their own field of endeavour.*
> (Lyn Lewis-Smith, CEO Business Events Sydney)

This book lays a challenge at the feet of the global business events industry to consider a reassessment of the way it thinks, the way it operates, the way it sees itself and the way it values and measures its achievements. It also asks governments and policy-makers to consider what broader legacies could be leveraged from business events to enrich the lives of those they govern. Finally, it seeks to inspire the students and new graduates who will one day be steering the industry. By expanding their understanding and vision of what's possible for business events today, there will be no limit to what can be achieved in the future.

Whilst many have contributed to the body of knowledge that has propelled this text forward, it is the first comprehensive and evidence-based attempt to argue that planning for, measuring and evaluating business events in terms of tourism expenditure alone is a limited perspective grounded in outdated thinking.

The book proposes that, by reconceptualising business events as important catalysts for change, the industry could position itself as one of the most important elements of social and economic development, driving knowledge generation and sharing, creating countless successful research partnerships, facilitating beneficial trade and political relations, recognising, stimulating and energising scholarly output, and encouraging development of emerging industry sectors and second-tier cities and economies.

All that is needed is a new perspective.

## A time to reflect

In 2020–2021 the business events industry, like many others, was challenged by the lockdowns, travel bans and social distancing restrictions associated with the COVID-19 pandemic. Mass gatherings of all kinds were cancelled, postponed or disrupted in destinations around the world and event industry professionals found themselves scrambling to implement new technologies capable of delivering virtual adaptations of the in-person meetings and conferences they had planned. Given the circumstances and the speed with which they had to pivot, what was achieved was remarkable, and at least some degree of connectivity was achieved. However, whilst most agree that virtual events are better than nothing, there is also broad consensus that the energy, inspiration and atmosphere generated by like-minded people gathering together physically at meetings and conferences was missed (Edwards *et al.*, 2020).

As vaccines and greater immunity slow the spread of the virus, face-to-face business events will almost certainly resume because, as humans and as professionals, we are drawn to gather with our peers. The small talk with strangers that goes on as delegates wait in the registration queue will again lead to friendships, networks and partnerships. The casual conversations over lunch will again spark inspirational ideas and solutions to problems. And the serendipitous introductions made during pre-dinner drinks will once again lead to funding opportunities that talented researchers had never dreamed possible.

The networking and interaction that occurs at in-person business events is very hard to achieve online. However, virtual events do have their merits. They are generally more accessible in terms of mobility, finance and time. They don't involve jet lag for long-haul passengers and they offer easier access for those not able or willing to travel. The carbon footprint of virtual attendance is better for the planet, registration fees are lower and there is no need to pay for accommodation. Essentially, *more* people may be able to attend *more* events *more* often if they are offered virtually. Because of this, the world will emerge from the pandemic period with a 'new normal' for business events with many more virtual and hybrid options being available to delegates than in the past.

However, it may well be that the new 'normal' for such events will mean fewer hotel rooms and cups of coffee are consumed by delegates. This could prove challenging for an industry which has etched out its worth and value using such measures.

So, as the events industry takes stock of what it has learned in recent times and reconfigures itself to account for this new normal, it may also be timely to consider whether there is need for an entirely different approach to the way business events are conceptualised, leveraged, planned, delivered, studied and evaluated in the future.

The impetus for this book began over a decade ago, when the authors began investigating whether conferences and business events created benefits and legacies beyond those related to tourism (Foley *et al.*, 2010; Foley *et al.*, 2013). Early studies suggested they did, and research tools were developed to identify just what those benefits were, how frequently they occurred and what impacts they generated (Edwards *et al.*, 2011; Foley *et al.*, 2014; Edwards *et al.*, 2016; Edwards *et al.*, 2017). The product of their research, along with the works of others (for example, Jago & Deery, 2012; Hansen *et al.*, 2020), is prompting a shift in attitudes, as industry leaders recognise that defining the success of a business event in terms of simply a tourism contribution limits its potential value. As competition for public funding, potential delegates' time and other resources increases, the higher the value proposition a business event can demonstrate, the more likely it will be sustainable and successful over the long term.

It also follows that, by identifying, planning for, and achieving the broadest possible range of social, economic, environmental, political and knowledge-based benefits from business events, the more sustainable, resilient and successful the industry itself will become.

## The tourism/events partnership

For decades, business event professionals and governments have operated under the narrow assumption that event legacies are best measured by their tourism contribution (Dwyer *et al.*, 2016; Foley *et al.*, 2013). But how did the value and worth of business events become so inextricably tied to tourism generation and expenditure? In very simple terms, it was a matter of 'right place; right time'.

Tourism (travelling) and events (gatherings which connect people and communities) have always been important aspects of human behaviour. However, it wasn't until the latter decades of the 20th century, with growth in both the public provision and commercialisation of leisure, that a small number of Western universities began offering relevant diplomas and subsequently degrees. These were initially leisure studies degrees, which covered multiple areas (recreation, sport, tourism, hospitality and events). This meant graduates had knowledge and professional skills in each of these areas, and understood the linkages and relationships between them. Tourism and then events soon emerged as scholarly fields in their own right at a time when both industries were becoming important to destination economies.

However, there were those who challenged the worth, legitimacy and long-term career potential that would come from these new industries. They questioned why one needed a degree to be able to manage tourism and events and whether the contributions graduates could make to society warranted the tertiary education resources being invested in them. Given these challenges, the developing industries recognised the importance of being able to justify their existence. Those in the field knew there were countless social and cultural benefits that involvement in event and tourism activities could bring, but what was needed in the neo-liberal managerial environment that characterised the era was measurable, tangible data that could be used to defend the legitimacy of what they did.

In terms of events, one important measurable and indisputable characteristic was the contribution they made to destinations in terms of injecting new money into economies as a result of direct tourism spending. Consequently, a somewhat symbiotic relationship between events and the tourism industry evolved, where events were seen as an effective strategy for boosting visitation and tourism expenditure at host destinations, and tourism expenditure at host destinations was used both as a way of garnering local support and investment in events, and as a tangible way of measuring the event's successfulness. In this way, the events industry became very much linked with the visitor economy, and management and support for events became the responsibility of destination managers and ministers with tourism portfolios. Business events, in particular, brought high-spending tourists and were seen as being particularly beneficial for local economies. Decision-makers, particularly those in government, like hard numbers, and measuring increases in expenditure associated with an event is relatively easy to report in hard numbers. Methodologies for measuring tourist expenditure were developed, adapted, critiqued and modified over time and the tourism-related economic benefits a business event could generate soon became the most important focus of both bid planning and post-event evaluation.

However, the premise of this book is that the tourism contribution for host destinations is merely the tip of the iceberg when compared with the other benefits that business events generate. It is also argued that this narrow focus on short-term tourism expenditure has seriously limited our understanding of the many rich legacies such events have the capacity to deliver, and the development of research, planning and management tools and strategies to identify, track and maximise those legacies.

Most governments perpetuate the undervaluing of business events with reporting requirements that are largely limited to metrics around the tourism contribution generated by congresses, conferences, symposiums, exhibitions and trade shows, and the convention bureaux and venues that support them. Most often located in the tourism portfolios of government ministries, business event bureaux and convention centres have focused their energies and resources for far too long on maximising the tourist dollars generated by business events, while ignoring the significant scientific, social, research and long-term economic value on offer; value that has the potential to benefit communities around the world by directly driving development, creativity, partnerships and innovation.

But with the business event industry now well established, it is time for event bidding, planning, execution and evaluation to go well beyond this rationale. Over recent years, event industry leaders have begun recognising the broader impacts and legacies events can generate, well beyond visitor expenditure (see Business Events Sydney, 2021; ICC Sydney, 2021; JMIC, 2021; The Iceberg, 2021). Leading researchers, including the contributors to this book, have investigated, measured and reported on the extensive legacies of business events beyond the tourism spend.

## Looking ahead

This book seeks to contribute to the existing research on business event outcomes, and to help inform discussion about the future of the industry. It includes abridged versions of the case studies listed in Table 1.1 below. These case studies, drawn from destinations around the world, were developed in order to capture data on the outcomes and legacies (in addition to tourism) generated by selected business events.

Each chapter introduces one case study and provides theoretical discussion on some of the relevant issues raised. The case studies are all very different in nature and are drawn from different types of business events; some large and some small. The studies relate to conferences, trade shows and meetings at universities from around the world. Some are based in capital or major cities (London, Sydney, Copenhagen, Amsterdam, Geneva) and others in regional destinations (Jonkoping, Sweden; Kuching, Malaysia; Punta del Este, Uruguay). Some of the perspectives presented are those of destination managers and convention bureau staff, some are from delegates and some belong to conference hosts/organisers. The case studies are presented in no particular order. Rather we invite the reader to start with whichever chapter is of most interest to them.

**Chapter 2** explores the concept of 'value', based around a case study of 12 international conferences and congresses held in Sydney. A broad cross-section of event types is included in the study, which sought to establish delegate perspectives on the benefits they perceived as a result of attending an event. The types of outcomes investigated included such things as whether delegates felt:

- they had been exposed to new ideas, knowledge, techniques and technologies;
- business collaborations and networks had been created;
- global research partnerships had been formed;
- the local sector's expertise had been showcased; and
- the international profile of Sydney had been raised.

The theoretical discussion which follows the case study focuses on the importance of such outcomes, their crucial role in healthy, thriving economies and the potential for business events to be seen as critical catalysts in the generation of these outcomes. It considers the virtuous dynamic of conference legacies and the processes of knowledge transfer; important theoretical perspectives that help explain how business events can maximise their legacy contributions to the knowledge economy.

In **Chapter 3** we consider the capacity for business events (and even the business events industry itself) to stimulate economic and social development, particularly for second tier or less developed regions. The case study looks specifically at the 55th International Conference and Convention Association (ICCA) Congress which was held in Kuching, Malaysia in 2016. From the outset, the bid to host the event was designed to support Malaysia's former Prime Minister Mahathir Mohamad's goal of seeing Malaysia recognised as a developed economy by 2020. It was hoped that the conference, which is the peak event for business event professionals across the globe, would not only stimulate the local economy as a result of such an important event being held in Kuching, but that it would showcase the city and Malaysia as a whole to the business event industry. Attracting business events to the destination was considered a crucial step toward becoming a fully developed economy, and legacies around upskilling and educating Malaysian event professionals were planned for and delivered.

But even more importantly, this case study refers to the significance of the inter-organisational collaborations that were established between various entities through the planning and execution of this major business event, and the beneficial outcomes that resulted from such relationships. In fact, Malaysia went to great lengths to ensure they generated long-term value from this event, well beyond tourism expenditure. Event industry professionals were able to establish links with government decision and policy makers and there were high levels of community engagement. They ensured important collaborations were established, and that social and ecological projects of significance to the local community (such as programs providing books for children and charities caring for local orangutans) were supported. Finally, they established engagement and linkages between ICCA Malaysia members who had previously worked somewhat in isolation.

The case study provides some great insights into how developing regions can use business events to plan for and achieve far-reaching and important legacies, beyond simple tourism expenditure. The chapter also outlines the many ways in which business events contribute to the United Nations Sustainable Development Goals.

But as our next chapter demonstrates, it's not just second tier cities that can strategically leverage themselves to generate promotional and long-term economic advantages and benefits business events offer. The focus of the case study in **Chapter 4** is London Tech Week, Europe's largest festival of live tech events. This week-long festival sees hundreds of different events conducted in various venues across the city, many run by tech-savvy, entrepreneurial local start-ups. London & Partners, the Mayor of London's official promotional agency, oversees the organisation, promotion and evaluation of the event which is designed to support and shape the future of the local tech industry. It brings new technological talent, knowledge and investment to London, connects the entire local tech ecosystem, and positions the city as the heart of technological development in Europe.

Whether it's smart homes, new medical techniques, advanced communications or artificial intelligence – our reliance on technology is increasing rapidly. This festival essentially showcases the best of London's technological talent and provides them with opportunities to connect with international financial development groups, business strategists, and industry leaders and decision-makers. It also promotes London as the first choice for overseas entities looking to expand into Europe or the United Kingdom, which creates significant long-term economic and job growth for the region.

The case study includes some useful insights and approaches to measurement which have broad applicability for those seeking to create similar legacy outcomes from their events. It also shows that by being clear about one's desired outcomes and having strategies in place to appropriately monitor, deliver and evaluate those outcomes, it is certainly possible to measure benefits beyond those immediately generated as a result of tourism.

The acquisition, formation and application of new knowledge, and the potential for conferences to become integral components of the process of global knowledge generation and sharing, is the focus of **Chapter 5**. When appropriately planned and delivered, conferences act as temporary contexts for learning and interaction which can trigger intense knowledge exchange mechanisms. This chapter makes reference to things like the spiral of knowledge (as it is upwardly shared) and the Four Es of learning (exposure, experience, expertise and embedding), and how such concepts can be used to guide conference design to ensure greater beneficial outcomes for delegates, their immediate communities of practice and their broader professional field.

The case study presented in **Chapter 5** demonstrates how some of these concepts were at play during the 2017 Euro Heart Care conference held in Jönköping, Sweden. This international conference is a major driver of research, ideas and collaboration in cardiac health care where knowledge-sharing and networking is actively encouraged by the organisers in order to maximise the potential of the learning opportunities generated. Delegates and organisers were interviewed after the event to gain an understanding of the benefits they felt it had generated. Many interviewees referred to the way they had acquired knowledge at the conference and then passed that knowledge on to those they worked alongside in the field (spirals). Others noted benefits around recognition of professional achievements, building of friendships, collaborations and networks, and the opportunities the event posed for professional development.

This case study gives food for thought in terms of just how far conferences can go in becoming catalysts for substantial professional growth, global connectivity and progression of fields of knowledge, and as stimulus for intense knowledge generation. It also demonstrates how legacy outcomes can transcend the short-lived experience of the business event itself.

In **Chapter 6** the focus shifts to sustainability. It looks at how those managing or working in the events industry can use their significant purchasing power to incorporate sustainable practices, to not only contribute to the viability of local products, producers, services and communities, but to give themselves a competitive edge. The case study looks at the Feeding Your Performance program, operating out of Sydney's International Convention Centre (ICC Sydney). The ICC

Sydney was opened in 2016 and is the largest integrated exhibition, convention and entertainment venue in Australia. With the United Nations declaring 2017 the International Year of Sustainable Tourism for Development, management chose to incorporate sustainable practices into their daily operations, establishing the Feeding Your Performance program and embedding the philosophy into all areas of daily operation.

The case study gives some very real examples of how the program operates and takes a detailed look at the range of beneficial outcomes (for the community *and* for ICC Sydney itself) which have resulted. These outcomes have been even more extensive than first anticipated. Apart from the supportive networks they've created with local producers and the knowledge that ICC Sydney is making a positive contribution towards the health of the regional economy and the environment, less altruistic benefits have also emerged. International and interstate clients are excited to be treated to local produce, and the high quality of the food and its presentation has a powerful impact on delegate performance and their perceptions of the event experience as a whole. Employees are more satisfied knowing that what they're doing is really making a difference, and there are improved links between ICC Sydney and the local agricultural sector in New South Wales. Finally, ICC Sydney has been able to extend their relationships and influence with important stakeholders, including government ministers who recognise the important contribution ICC Sydney is making to the regional and local economies under their care.

The chapter goes on to discuss broader concepts such as the increasing propensity of customers to seek out products and services that are procured, produced and distributed ethically and sustainably. Whilst competitive pricing remains important, most are willing to pay a little more to support those entities practising corporate social responsibility (CSR). Things like sourcing locally, supporting small business, local job growth and regional development, and ethical practices are seen to be given a place value as part of the buying decision. There is also growing awareness in the community of animal welfare and fair-trade issues, environmental impacts associated with industrial mass food production and global supply chains (food miles), and a focus on the origins of food and food safety standards. With these things in mind, the capacity for hospitality venues, convention centres and business events to incorporate sustainable practices into their operations is becoming increasingly important.

Accordingly, Feeding Your Performance has become a cornerstone of the ICC Sydney marketing strategy and a point of difference upon which they promote the facility and its events.

In the theoretical discussion of the case study, we look at the benefits business event professionals might expect to realise by incorporating CSR into their operations, such as heightened employee morale, improved competitive advantage, better management of risk, increased market share and profit, enhanced reputation and social image, stronger value chains as a result of responsible sourcing, cost reductions and access to global markets.

This chapter provides an excellent platform from which industry players can extend their concept of 'what's possible' and incorporate sustainable practices, not only to differentiate themselves from the competition but to become a catalyst for positive change by promoting socially inclusive and sustainable economic growth.

The focus of **Chapter 7** is on the relationship between academics and conferences. Academic conferences make up a significant proportion of all business events held throughout the world each year. Similarly, academic delegates and presenters make up a significant proportion of overall conference participation. Academic researchers are often asked to host or support a bid for a conference, and the largest audience for the consumption of conference products (reports, papers, etc.) is usually made up of academics.

Yet little research or investigation has been done into what drives academics to attend, present at, or host such events, or how successfully their expectations of outcomes, benefit or legacy have been met. There are certain traditions and conventions which help shape the way such business events are planned, delivered and evaluated. There are also some rather unique reasons academics are motivated to attend. It follows, then, that if the potential value such events hold is to be maximised by event professionals it is important that a deeper understanding of the drivers behind academics' attendance, involvement and support are more fully understood.

The chapter first provides context on the evolution of the academic conference environment and raises a number of important issues in event delivery facing this sector. The case study used in the chapter to illustrate some of these issues was drawn from a study of six international research conferences and two failed attempts to host a conference in four European countries. Semi-structured interviews were conducted with a range of stakeholders, including lead researchers, the scientific societies behind the conferences, representatives of universities hosting the conferences and event professionals from local convention bureaux, professional conference organisers (PCOs) and venues where conferences were held. The case study gives a great summary of the range of possible benefits academics perceived were able to be attained as a result of conference attendance or hosting, and also some barriers which prevented them from either proceeding with plans to host a conference or from attending a conference. In summary, academics attend conferences to develop and expand networks, initiate collaborations, gain recognition, further their careers, build global communities of friendship with like-minded peers, learn about developments in the field, promote and disseminate their research and ultimately to make contributions to the broader community

This important chapter will help event professionals who work with academics in any capacity to better understand what motivates them and how to provide a more tailored and successful event experience. Event professionals need to discover why academics attend (or don't attend) conferences, what motivates them to present their work at business events, and how hosting an academic conference can create long-term career benefits for those willing to embark upon the task, as well as for their employing institutions.

The theoretical discussion at the end of the chapter seeks to link the perceptions and experiences detailed in the case study with an understanding of how social power (capital) and recognition are important drivers of academic behaviour. It looks at the academic credibility cycle and considers the significance of the 'buzz' felt at a conference and its role in helping those who access it get more out of the experience. By linking the case study findings to these concepts, the chapter gives event professionals a solid platform from which to reimagine the academic conference, and to shape future events/bids/evaluations in ways that will more fully provide academics with the legacies they seek.

**Chapter 8** looks at the potential for business events to be used as a strategy of industry sector development. It investigates the Swiss Fintech (financial technology) Corner (SFC), which was a small part of the 2016 SIBOS conference. SIBOS (the Swift International Banking Operations Seminar) is a premier business event for the global financial community, which in 2016 was held in Geneva, Switzerland. Whilst Switzerland is known to have some advantages that see it well placed in terms of global financial centre competitiveness (centrality, high levels of security, stability, etc.), over recent years it had slipped in its rankings. This conference and, in particular, the Swiss Fintech Corner, was seen as a way to make up lost ground and to showcase the innovative financial technology applications Switzerland had to offer. Essentially, the Swiss Fintech Corner aimed to link the local fintech start-ups with international players, raise Geneva's profile as a fintech hub, and increase awareness of the importance of fintech start-ups in Geneva amongst public and private authorities.

The case study makes use of interviews conducted post-event with stakeholders, policy makers and the individual start-ups involved. The study also utilises demographic data on SIBOS delegates and social media metrics to give greater context. Whilst the study attempted to focus specifically on the SFC, it does include significant material on the wider benefits to innovation, industry, and Geneva and Switzerland's image as a financial centre that resulted from hosting the wider SIBOS event. Importantly, the study notes the difficulty in measuring such outcomes in the absence of pre-event planning (for legacy or evaluation) and appropriate tools.

The theoretical linkages are around the way networks are shaped (and shape) events, and how they have the potential to create local distinctiveness and innovation. Start-ups play an important role in innovation, but they often have problems linking with the 'big end of town'. When events are strategically used to connect emerging talent and ideas with business clients, investment groups and government, they are seen to be supporting the innovation needs of local economies and making important contributions to thriving economies.

Finally, in **Chapter 9** the capacity for business events to be used as drivers of trade, foreign investment, political partnerships and infrastructure development is considered. The China-LAC Business Summits are designed to establish or maintain mutually beneficial business partnerships between government entities, investors and business operators from China, Latin America and Caribbean countries. The 11th summit was held in 2017 in Punta del Este, Uruguay, and acted as a catalyst for regional economic growth and development. Those interviewed for the case study indicated their expectations were not only met, but were exceeded, in terms of attendance at the event and the number of partnerships and investment networks generated. Once again, the legacy concepts raised in the study provide a platform upon which event industry professionals and researchers can build. In particular, there is evidence of the capacity for such events to improve trade and innovation pathways and facilitate opportunities between the countries involved. The China-LAC summit created strategic commercial ties between important investors (from China) and LAC countries seeking funding for business and infrastructure development. The study measured media coverage to evaluate return on investment by monetising the free marketing exposure they received as if it had been paid advertising.

And whilst the summit itself aimed at these broader economic outcomes and legacies, there were strategic, long-term economic objectives driving Punta del Este's bid as a host destination. Punta del Este is a highly seasonal tourism destination, regarded somewhat as a playground for the rich. Poorer residents are pushed to the outskirts of the city and, once the summer tourism boom ends, business activity in the heart of the city significantly decreases. To counter this, Uruguay's government hoped that hosting a successful summit would position Punta del Este as a meetings industry destination of the future, and that local event professionals would gain valuable skills via the hosting process.

Overall, as a result of the strategic partnerships established, this single event was used to generate multiple legacies with long-lasting implications for the host destination and the broader national economies of the countries involved.

The chapter's theoretical discussion looks more deeply at how high-profile events can be instrumental in creating significant positive change in international relations and trade. It is argued that business events have the capacity to contribute to important government policies through the development and growth of 'soft diplomacy' strategies at state, national and international levels. China's use of soft power (public and cultural diplomacy and relations) in its economic development and reduction of poverty has been both strategic and highly successful over the past few decades, and business events have played an important role in providing opportunities for such strategies to be implemented.

The case study is a good example of how business events can be used to leverage soft power for political, social, trade and economic benefit. Those event industry practitioners who shift their perspective to determine how best to leverage, deliver and measure such benefits will make an important contribution to the value and long-term viability of their industry.

# In summary

This book is written for event management scholars, business event professionals, and the many event management students who will make their mark in the world as responsible global citizens during their successful careers; driving legacy, innovation and social change in the highly valuable business events industry.

The diversity of perspectives presented in this book provide rich insights into just a few of the many important impacts *beyond* tourism-related economic stimulus that business events can bring. These new perspectives signal a maturation of the business events industry that will shape future attitudes and directions. Once greater understanding of this broad range of impacts is obtained, those working in the industry will be better equipped to sustainably leverage events in ways which maximise potential benefits for destinations, delegates, hosts, local businesses/communities, knowledge economies and the global community as a whole.

Perhaps of even greater importance is the potential such a shift in perspective and practice could bring to the industry in the future. If the full extent of the types of event legacies discussed in this text were able to be identified, measured, planned for and achieved, the business events industry would have the capacity to position itself as an integral component of sustainable social, economic and global development.

## A note about our contributors

The case studies in this book are abridged versions of cases submitted as part of a research project conducted by the University of Technology Sydney on behalf of the Joint Meetings International Council (JMIC). JMIC is an organisation that represents the interests of the major international business meetings industry associations and its membership comprises the global leaders in the business events industry. JMIC is currently focused on "*documenting and communicating the diverse values of the business events industry to broader community and governmental audiences in order to generate greater appreciation for its diverse roles in supporting global economic, academic, professional and social development*" (JMIC, 2021).

Global leaders in the business events industry are aware that the legacies arising from conferences, conventions, exhibitions, trade shows and related business events are important in achieving professional, industry, academic, community and economic objectives. Being able to demonstrate the extent of these benefits is central to ensuring future business and to creating goodwill for the industry among communities and governments. The JMIC case study initiative (Foley *et al.*, 2019b) was a coordinated, global effort to measure and document these legacies in ways that could be used to advocate on behalf of the industry.

The original cases were written by business event industry professionals and academics from Asia, Australia, Europe and South America. These cases and their authors are listed below.

**Table 1.1:** Case studies referred to in the book

| |
|---|
| Thomas Trøst Hansen. *International Research Conferences: The Academic Impact*, Aalborg University and Wonderful Copenhagen (see Hansen, 2018). |
| Andrew Gambina, Johanna Hagström and André Petzold. *Knowledge Sharing and Organisational Development Through the 2017 Euroheartcare Conference*. Jönköping University and Destination Jönköping (see Gambina *et al.*, 2018). |
| Miriam Scaglione and Simone Dimitriou. *Zoom-in on the first-ever Swiss Fintech Corner*. University of Applied Sciences and Arts Western Switzerland Valais (see Scaglione & Dimitriou, 2018). |
| Manisa Mohamed Nor. *The Sustainability of the Business Events Industry in Malaysia: Leveraging Inter-Organisational Collaboration for the 55th ICCA Congress*. Malaysia Convention & Exhibition Bureau and MyCEB Congress Malaysia (see Mohamed Nor, 2017). |
| Gaining Edge. *Legacies from ICCA Congress 2016 Kuching SARAWAK: How Hosting ICCA 2016 Impacted Our Future*. Business Events Sarawak and Borneo Convention Centre (see Gaining Edge, 2020). |
| Amandine Thiebault. *London Tech Week*. London & Partners (see Thiebault, 2018). |
| Hilary du Cros, Carmel Foley, Deborah Edwards and Anja Hergesell. *Catalysts for Thriving Economies*. University of Technology Sydney and Business Events Sydney (see du Cross *et al.*, 2017). |
| Carmel Foley, Deborah Edwards, Bron Harrison and Anja Hergesell. *ICC Sydney: Feeding Your Performance*. University of Technology Sydney and ICC Sydney (see Foley *et al.*, 2019a). |
| María José Alvez, Irene Sinigaglia, Alanis Lourdes and Gustavo Perrier. *11th China-LAC Business Summit 2018*. Punta del Este Convention & Exhibition Center, Uruguay XXI, Grupo Elis and Arquitectura Promocional (see Alvez *et al.*, 2019). |

## Acknowledgement

The initial case study project run by the University of Technology Sydney on behalf of the Joint Meetings International Council (JMIC) was supported by an academic advisory board – Professor Leo Jago, Professor Larry Dwyer, Associate Professor Karin Weber, Professor Betty Lou, Professor Markus Grosse-Ophoff and Professor Paul Verdin. We extend our sincere thanks to these esteemed academics for their contribution in assessing the case studies.

## References

Alvez, M., Sinigaglia, I., Lourdes, A. and Perrier, G. (2019) 11th China-LAC Business Summit 2018. Punta del Este Convention & Exhibition Center, Uruguay XXI and Arquitectura Promocional. https://www.themeetingsindustry.org/web/wp-content/uploads/JMIC_Case-Study_Punta-del-Este_CHINA-LAC-Business-Summit.pdf

Business Events Sydney (2021) Research. https://www.besydney.com.au/about-us/resources/publications/besydney-research/

du Cros, H., Foley, C., Edwards, D. and Hergesell, A. (2017) *Catalysts for thriving economies*, University of Technology Sydney and Business Events Sydney. https://www.themeetingsindustry.org/web/wp-content/uploads/JMIC_Case-Study_BE-Sydney_Catalysts-for-thriving-economies-17.5.pdf

Dwyer, L., Jago, L. and Forsyth, P. (2016) Economic evaluation of special events: Reconciling economic impact and cost–benefit analysis, *Scandinavian Journal of Hospitality and Tourism*, **16** (2), 115-129, doi: 10.1080/15022250.2015.1116404.

Edwards, D., Foley, C. and Hergessel, A. (2020) *Conferences in a Covid-19 world*, report submitted to Business Events Sydney.

Edwards, D., Foley, C. and Hergesell, A. (2016) *Conferences: Catalysts for Thriving Economies*, Sydney: Business Events Sydney, viewed 14 December 2016. https://www.businesseventssydney.com.au/media/2070/catalysts-for-thriving-economies-research.pdf.

Edwards, D., Foley, C. and Malone, C. (2017) *The Power of Conferences*, UTS E-Press. https://utsepress.lib.uts.edu.au/site/books/m/10.5130/978-0-6481242-0-7/

Edwards, D., Foley, C. and Schlenker, K. (2011) *Beyond Tourism Benefits: Measuring the Social Legacies of Business Events*, Sydney: Business Events Sydney, viewed 14 December 2016. https://www.businesseventssydney.com.au/media/1366/beyond-tourism-benefits.pdf.

Foley, C., Edwards, D., Harrison, B. and Hergesell, A. (2019a) ICC Sydney: Feeding Your Performance, University of Technology Sydney and ICC Sydney. https://www.themeetingsindustry.org/web/wp-content/uploads/JMIC_Case-Study_ICC-Sydney_Feeding-your-performance_Y2_18.9.pdf

Foley, C., Edwards, D. Jasovska, P. and Hergesell, A. (2019b) *Business event legacies*, report submitted to JMIC. https://www.themeetingsindustry.org/news-media/media-releases/

Foley, C., Edwards, D. and Schlenker, K. (2014) Business events and friendship: leveraging the sociable legacies, *Event Management: An International Journal*, **18**, 53-64.

Foley, C., Schlenker, K., Edwards, D. and Hayllar, B. (2010) *A Scoping Study of Business Events: Beyond Tourism Benefits*. Sydney: Business Events Sydney.

Foley, C., Schlenker, K., Edwards, D. and Lewis-Smith, L. (2013) Determining business event legacies beyond the tourism spend: An Australian case study approach, *Event Management: An International Journal*, **17**, 311-322.

Gaining Edge (2020) *Legacies from ICCA Congress 2016, Kuching, Sarawak: How hosting ICCA 2016 impacted our future*, report prepared for Business Events Sarawak and Borneo Convention Centre.

Gambina, A., Hagström, J. and Petzold, A. (2018) *Knowledge Sharing and Organisational Development through the 2017 Euroheartcare Conference*, Jönköping University and Destination Jönköping. https://www.themeetingsindustry.org/web/wp-content/uploads/JMIC-Case-Study_Jonkoping_Euroheartcare-2017_18.6.pdf

Hansen, T. (2018) International Research Conferences: The Academic Impact. Aalborg: University and Wonderful Copenhagen. https://www.themeetingsindustry.org/value-proposition/case-studies/.

Hansen, T., Pedersen, D. and Foley, C. (2020) Academic events: An empirically grounded typology and their academic impact, *Event Management*, **24**, 481-497.

ICC Sydney (2021) Legacy programs. https://www.iccsydney.com.au/Venue/About-ICC-Sydney/CSR/Legacy-Program

Jago, L. and Deery, M (2012) *A theoretical framework for business events: The role of business events in innovation and knowledge creation, CAUTHE conference proceedings*, La Trobe University, Melbourne, Victoria, pp. 294–300.

JMIC (2021) Case studies. https://www.themeetingsindustry.org/value-proposition/case-studies/

Mohamed Nor, M. (2017) The sustainability of the business events industry in Malaysia: Leveraging inter-organisational collaboration for the 55th ICCA Congress, Malaysia Convention & Exhibition Bureau. https://www.themeetingsindustry.org/web/wp-content/uploads/JMIC_Case-Study-Program_MyCEB_ICCA-congress-17.5.pdf

Scaglione, M. and Dimitriou, S. (2018) Zoom-in on the first-ever Swiss Fintech Corner, University of Applied Sciences and Arts Western Switzerland. Valais. https://www.themeetingsindustry.org/web/wp-content/uploads/JMIC_Case-Study-Program_Geneva_Swiss-fintech-at-Sibos-17.5.pdf

The Iceberg (2021) Legacies of business events, Joint Meetings Industry Council. https://www.the-iceberg.org/

Thiebault, A. (2018) London Tech Week. London & Partners. https://www.themeetingsindustry.org/web/wp-content/uploads/JMIC_Case-Study_London-and-Partners_London-Tech-Week-17.12.pdf

# 2 Conferences as Catalysts for Thriving Economies

> *Knowledge creation is increasingly dependent on collaboration, which relies on the camaraderie, trust and relationships formed through face-to-face contact. Business events provide the platforms for this physical in-person interaction and the foundation for the networks that then drive collaboration and innovation.*
>
> (Lyn Lewis-Smith, CEO Business Events Sydney)

This chapter explores the changing perceptions of the value of business events. As discussed in Chapter 1, formal industries and tertiary education courses in tourism and event management emerged in the latter decades of the 20th century. During that period, Western economies were being influenced by economic rationalism, prompting some to question the contributions such industries could make to the economic viability of society, expansion of business and job creation. One important argument in support of the development of these industries, therefore, was the economic stimulus they could provide for local economies by way of tourism expenditure. A symbiotic relationship developed between the events and tourism industries, with events being seen as an effective way of generating additional tourism activity, and tourism expenditure being used to justify and evaluate the successfulness of an event. This was particularly so in terms of business events, given that visitors to business events have higher daily expenditures compared to those of all other tourists (Foley *et al.*, 2014a; Foley *et al.*, 2014b), and as a result, the business event delegate came to be seen as the most desirable visitor in the world (Business Events Council of Australia, 2010).

But can business events provide value beyond their contribution to tourism expenditure?

The practice of rationalising, bidding for, investing in and evaluating the successfulness of business events based on the tourism spend they generated was both practical and understandable at a time when the events industry was etching out its place within neo-liberalist economies. However, with the events industry now well established, there were opportunities to look beyond the value provided as a result of direct tourism-related expenditure and to unlock the potential of business events to generate more significant and lasting economic and non-economic values.

New understandings of the value of business events beyond the tourism contribution began to emerge in the early decades of the 21st century. Business event scholars argued that business events should not be measured by direct financial contribution alone (Dwyer et al., 2000; Foley *et al.* 2014c; Edwards *et al.*, 2016; Edwards *et al.*, 2017), that there was a lack of recognition of the wider benefits of events (Carlsen *et al.*, 2001; Wood, 2009; Foley *et al.*, 2010), and that these non-tourism benefits had yet to be fully examined (Foley *et al.*, 2013). Many research, policy and business environments needed unique opportunities to showcase, construct and brainstorm new strategies, ideas and technologies in order to thrive, and things like live presentations and networking opportunities at conferences could create special impetus for developing new professional relationships and research collaborations that would survive well past the closing ceremony.



A number of industry reports (Business Events Industry Strategy Group, 2008; Foley et al., 2010; Jago & Deery, 2010; Teulan, 2010; Edwards et al., 2016) offered evidence-based examples of the broader economic benefits that business events could bring; however, very little peer-reviewed academic literature had examined indirect economic legacies and non-economic legacies. There was clearly a need to evaluate business events in more sophisticated ways that moved beyond the traditional economic impact measures (Dwyer et al., 2015). This case study (adapted from du Cross, Foley, Edwards & Hergesell, 2017) offers one exemplar in its examination of the value beyond the tourism contribution of a set of 12 international conferences held in Sydney, Australia.

## Case study: The value of conferences beyond tourism benefits

| | |
|---|---|
| **Business event(s)** | 12 international conferences of various sizes and sectors in 2014 and 2015 |
| **Location(s)** | Sydney, Australia |
| **Primary event legacy** | Business events facilitate interactions and collaborations that underpin innovation, economic development and societal change |
| **Additional legacies (beyond tourism)** | Knowledge expansion, industry innovation, networks, collaborations, social capital, community development, international relations, showcasing destination, attracting global talent |
| **Summary quote** | "Findings from this study demonstrate ... it is essential for governments to transition from valuing business events for their expenditure outputs to recognising them as transformative social network spaces." |
| **Lessons learned** | Outcomes need to be leveraged/planned in advance |
| **Future research** | Ways of documenting/measuring long-term impacts on idea generation, research collaborations, delegate careers, knowledge creation and industry development |

This case study was adapted from: du Cros, H., Foley, C., Edwards, D. and Hergesell, A. (2017)

# Introduction

This case study examines 12 international business events held in Sydney in 2014 and 2015. The events chosen were diverse in terms of both industry sector and size in order to ensure wide applicability of findings. Four events were from the cultural sector, three were medical, three were technical industry events, and there were two environmental congresses (see Table 2.1). Delegate numbers ranged from as few as 60 to as many as 6000.

Primary and secondary data sources (including questionnaires and personal interviews) were used to establish the demographic profiles of respondents and to garner their perspectives on whether certain legacies (beyond tourism benefits) had been generated as a result of the events.

The overall aims of the study were to determine whether such events (particularly international conferences) have the potential to act as catalysts for economies and to understand what can be done to maximise the extent to which business events are able to energise or reboot the professions and industry environments they support.

The case study was prepared by researchers from the University of Technology Sydney (UTS) Business School, who were commissioned by Business Events Sydney (BESydney) to investigate potential legacy outcomes arising from business events.

**Table 2.1:** Business events surveyed, dates held and industry sector

| Conference | Held | Number of Attendees | Industry Sector |
|---|---|---|---|
| World Harp Congress (WHC) | Jul 2014 | 604 | Cultural |
| World Congress on Mild Approaches in Assisted Reproduction (ISMAAR) | Sep 2014 | 168 | Medical |
| Polarised Neutrons in Condense Matter Investigations (PNCMI) | Sep 2014 | 60 | Technology |
| 4th International Congress of International Society for Cultural and Activity Research (ISCAR) | Oct 2014 | 342 | Cultural |
| World Green Infrastructure Congress (WGIC) | Oct 2014 | 371 | Environment |
| Congress of the International Society for Burn Injuries (ISBI) | Oct 2014 | 903 | Medical |
| International Conference on Walking and Liveable Communities (Walk21) | Oct 2014 | 502 | Cultural |
| Communicating the Museum (CTM) | Nov 2014 | 265 | Cultural |
| IUCN World Parks Congress (IUCN) | Nov 2014 | 6081 | Environment |
| 5th INS/ASSBI Pacific Rim Conference (PRC) | Jul 2015 | 672 | Medical |
| IEEE Multi-Conference on Systems and Control (MSC) | Sep 2015 | 317 | Technology |
| World Hydrogen Technologies Convention (WHTC) | Oct 2015 | 320 | Technology |

2

## Research methods

This case study was based on both qualitative and quantitative data obtained from delegates and organisers of the 12 business events listed in Table 2.1. Due to small sample sizes for some conferences, findings were enhanced by grouping conferences according to industry sector (Table 2.1, column 4).

### *Quantitative research*

On behalf of the research team, conference organisers emailed event delegates to notify them of the study. Questionnaires were then sent to those indicating they were willing to participate, using the online survey program Qualtrics. The questionnaire was designed to examine a range of demographics, the tourism aspects of business event travel, and legacy outcomes of business events. Given the breadth of the study, it was important that the survey was not too long. Previous research into beyond tourism benefits (BTB) of business events (Edwards *et al.*, 2011; Foley *et al.* 2013) had identified 45 possible legacy variables which business events might achieve. Factor analysis of those variables was undertaken to reduce the list to 13 linked legacy items (Table 2.2). These 13 items were then further grouped into four overarching legacy dimensions which informed survey design.

The questionnaire also sought to obtain demographic data on respondents and feedback on their overall level of satisfaction with the service and functional aspects of their visit to Sydney.

Using independent samples *t* tests and analysis of variance (ANOVA), legacy questions were analysed for any significant differences in terms of age, gender, delegate type, and place of origin. Statistical significance is based on SPSS-calculated independent samples *t* scores or ANOVA calculations. In statistical terms, differences termed 'significant' are considered unlikely to have been caused by chance alone. Unless otherwise stated, differences are noted at the $p < .01$ level.

**Table 2.2:** The dimensions of legacy

| Legacy dimension | Legacy items |
|---|---|
| Innovation | Ideas; Knowledge; Techniques; Technologies |
| Collaboration | Supported the development of global research and business collaboration |
| Sector development | Developed the knowledge and capabilities of early career delegates in Sydney<br>Developed professional practices that have enhanced outcomes for the community in Sydney<br>Improved the overall skills and ability of the sector in Sydney<br>Enabled the local sector to showcase their expertise to a global audience<br>Raised the international profile of Sydney |
| Attracting global talent | I would like to live and work/study in Sydney<br>I have applied for a position to work/study in Sydney<br>I know others who have relocated to Sydney |

## Qualitative research

In addition to the questionnaire survey, eight interviews were conducted over several weeks (during October and November 2016). Those interviewed represented a selection of stakeholders who indicated they were willing to comment on longer term legacies from the viewpoint of one to two years post-survey. Telephone interviews were conducted with the three event organisers (Professor William Ledger, University of New South Wales, Dr Hal Lee, Bragg Institute, NSW and Dr Irina Verenikina, University of Wollongong) and two researchers (Dr Ian Lilley, University of Queensland and Dr Celmara Pocock, University of Southern Queensland). A practitioner delegate (Robin Aitken, Sydney Area Manager, NSW National Parks and Wildlife Service) was interviewed by email. Finally, two staff from BESydney (Iain Stuart and Inga Davison) were interviewed by phone and by email, respectively. Comments from interview transcripts and emails were used to enhance understanding of survey results and provide deeper perspectives.

## Demographic data of respondents

There were 1110 completed surveys ($n = 1110$) and responses were grouped according to industry sector (Table 2.3).

**Table 2.3:** Number of usable responses by industry sector represented

| Industry sector | No. of useable responses |
|---|---|
| Cultural | 233 |
| Environment | 613 |
| Medical | 164 |
| Technology | 100 |

**Gender** Whilst there was a relatively equal spread of responses in terms of gender overall (56% from females and 44% from males), there were significant differences ($p < .001$) in the gender balance of responses received from particular events. For example, the majority of responses from the World Harp Congress were from females (93%), and the bulk of responses from the Multi-conference on Systems and Control were completed by males (92%). It was also noted that domestic respondents were more likely to be female and younger, and the majority had visited Sydney previously. International respondents, by contrast, were more likely to be older males and had travelled from North America or Europe to Sydney.

**Age** Seventy-five% of responses received were from delegates aged between 25 and 54 years. However, once again, responses from certain events indicated very different demographics, with the World Harp Congress, for example, having a significantly higher proportion of young respondents (up to 24 years of age) and older respondents (55–64 years of age). Significant differences were also noted in gender across all conferences, with females tending to be younger and males tending to be older.

**Place of origin** More than half of the respondents were from overseas (58%) and 23% were from interstate (states of Australia other than New South Wales). Of all international respondents, 96% came to Sydney because of the conference, staying an average of 4.6 days beyond the conference duration. Seventy-nine% of international respondents stated they intend to visit Sydney again.

Domestic respondents were more likely to be female (64%) and international respondents were more likely to be male (52%). The largest place of origin for interstate respondents was Victoria (38%), followed by Queensland (21%).

Practitioner delegates comprised 54% of all respondents followed by academic delegates (33%). There were also significant differences ($p < .001$) by age and gender, with exhibitors and academic delegates being younger (on average) than practitioner delegates, and academic delegates more likely to be male.

## Findings

As detailed in Table 2.3, the 13 legacy variables are grouped into four overarching legacy dimensions. They are:

- Innovation (new ideas, knowledge and techniques);
- Collaboration (supporting development of global research and business collaboration);
- Sector development (improved capacity of early career delegates, the local community and the relevant sector's overall performance in Sydney);
- Attracting global talent (bringing new talent to Sydney).

These four dimensions are used to present case study findings. Also presented are findings on how exhibitors perceived they had benefited from the event.

### *Innovation*

Overwhelmingly, respondents agreed the conference exposed them to new ideas and knowledge (see Figure 2.1).

There were, however, significant differences noted between certain demographics (ages; industry sectors) in terms of their perceptions of the evidence of benefits from the four innovation items (see Figure 2.1). Those differences were:

1. *Ideas*: Those aged between 40 and 44 years were less likely to agree that the conference exposed them to new ideas.
2. *Knowledge*: Respondents aged between 55 and 59 were less likely to agree. Those aged 50 to 54 years considered the benefit was still to be realised and those aged 30 to 34 were more likely to be unsure as to the benefit occurring. Respondents from the technology sector were more likely to agree that benefits were still to be realised, while their counterparts from the medical sector were confident benefits were realised. Respondents from the cultural sector, particularly music delegates, were the most unsure.

3. *Techniques*: Cultural sector respondents were less likely to agree. European respondents were most likely to disagree and feel unsure, while respondents from Asia were more likely to feel that benefits were still to be realised.
4. *Technologies*: Respondents from the environment sector were less likely to agree, while those from the cultural sector were more likely to disagree or to be unsure about this benefit. Academic respondents were more likely to disagree; however, exhibitor respondents were more likely to be unsure. Respondents from Europe were the most likely of all international attendees to disagree or be unsure.

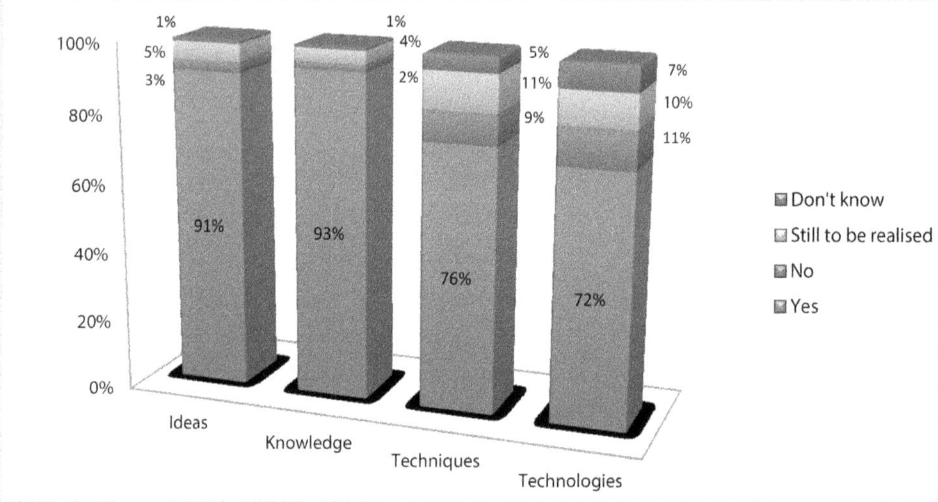


**Figure 2.1:** Percentage of respondents exposed to new ideas, knowledge, techniques and technologies

## *Collaboration*

There was strong support for the notion that conferences supported the development of global research and business collaborations, with 76% of respondents agreeing. Thirteen% also thought these benefits were still to be realised, suggesting a longer time frame may be required to establish such collaborations for some. The only significant difference noted was in respondents from the cultural sector, who were more likely to disagree and more likely to not know.

## *Sector development*

The majority of respondents (83%) agreed the conference enabled the local sector to showcase its expertise to a global audience, raising the international profile of Sydney (see Figure 2.2). This was supported by all event organisers interviewed, who stated they had received positive feedback from members about the conference location in Sydney (Ledger, Verenikina & Lee, personal communications, October 2016).

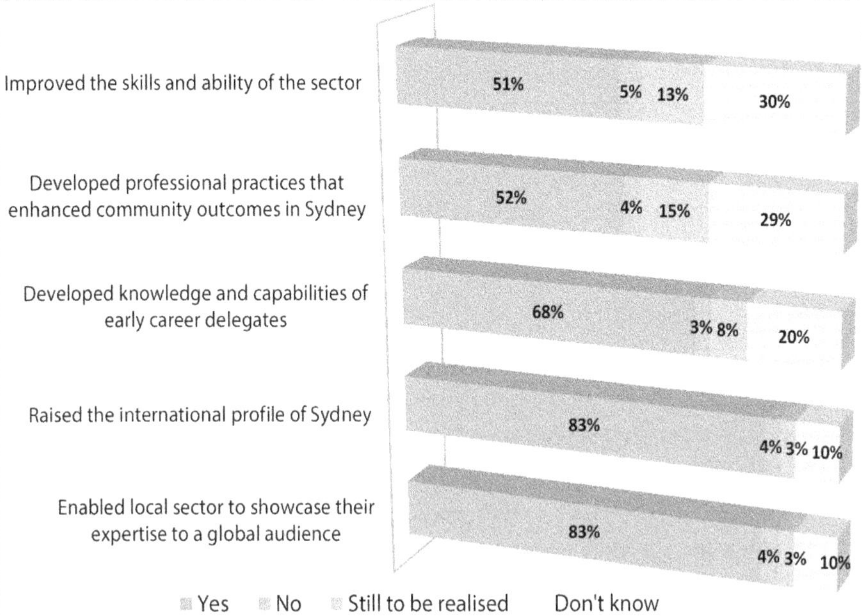


**Figure 2.2:** Sector development

Significant differences (between demographic groups) were once again noted, however, on the following items:

1. *Knowledge and capabilities of early career delegates in Sydney*: Respondents from the environment sector events were more likely to feel this benefit is still to be realised, while those from the medical sector were more likely to see the benefits already being realised. Those aged 25 to 29 years were more likely to disagree such benefits had occurred, and those aged 25 to 34 years considered the benefits were yet to be realised.
2. *Professional practice that enhanced outcomes for the community in Sydney*: Asian respondents were more likely to agree, while European respondents were more likely to respond that they did not know. Medical sector respondents were more likely to agree, and consider the benefits were realised at the time of the conference.
3. *Overall skills and ability of the local sector*: Music and medical sector respondents were more likely to agree than those from the technology and environment sectors. Respondents from Asia were less likely to respond 'did not know', and those from Europe were less likely to agree and more likely to respond that they did not know.

Respondents also stated they felt the conference they attended allowed:

1. *The local sector to showcase their expertise to a global audience* – with most European respondents agreeing with this statement. Many from New Zealand/Oceania (and some from Europe) stated they did not know. Respondents from the technology

sector were more likely to agree than those from other sectors that this benefit was still to be realised.

2. *The international profile of Sydney to be enhanced* – with respondents from the technology sector being more likely to believe this benefit was still to be realised, and those from the cultural sector more likely to not know. A significant difference was found in the study of age groups, with early career academics (25 to 29) more likely to consider this benefit was still to be realised, while those aged 40 to 44 were more likely to be unsure.

### *Attracting global talent*

Delegates felt conferences held in Sydney serve as an attractor for global talent, with 7% stating they had applied for a position to work/study in Sydney as a result of attending the conference, and 41% stating they would like to live and work/study in Sydney.

Significant differences by age were noted in relation to the impact of the conference on delegates wanting or applying to live, work and/or study in Sydney. Of the respondents who said they had applied for a position to work/study in Sydney as a result of the conference, those aged 18 to 24 and 30 to 34 were more likely to agree with the statement, while respondents aged 50 to 54 and 60 to 64 were less likely to agree.

### *Exhibitors' perceived benefits*

Twenty-five per cent of exhibitors said they had increased sales and 41% made international sales contacts as a result of their attendance at the business event.

## Discussion of findings

In discussing these findings, the qualitative data obtained from event organisers and willing delegates via personal interviews is used to illustrate ways in which survey data relates to business event outcomes.

### *Key legacies*

Aligning with earlier research, the findings suggest the career stage of a delegate may be directly relevant to the legacy outcomes obtained at an individual level (Mata *et al.*, 2010; Edwards *et al.*, 2011). For example, early and mid-career researchers are more likely to be stimulated by new ideas and knowledge than later career researchers. It may also suggest conference attendance is an important step for career development (Vega & Connell, 2007).

An interesting distinction between types of conferences emerged from the data. Delegates from the technology sector were more likely than their medical counterparts to think there were benefits to be realised well beyond the conference. This result may indicate that innovation in the medical sector is more immediate and tangible in application than is possible or practicable in the technology sector. Previous research on medical sector conferences "found that experimental research findings are important to the [medical] practitioner due to their greater relevance for long range applicability to practice" (Clickner *et al.*, 1998: 12) as well as their application for practice (Christiaens *et al.*, 2008). Additionally, delegates attending music conferences reported the enhancement of skills and capability in the sector as clear conference outcomes. This suggests that music conferences also have a strong orientation to practice.

There are other benefits which may have been obtained by delegates attending the cultural conferences which have not been fully captured in this research. These may include quality of life enhancements, the distribution of cultural ideas, and effects on the extra-aesthetic environment in which cultural activities are performed and received. Further research is required to investigate these sorts of distinctions. It does appear certain, however, that benefits derived from business events are cumulative, occurring not only during the conference but also before and after the event. The findings continue to support previous studies (Foley *et al.*, 2013; Foley *et al.*, 2014c), which found some of the most significant social outcomes from international conferences occurred years after the event was held.

In addition, the benefits identified have been extended to a diverse range of stakeholders, from individual delegates through to the broader community. Delegates gained new knowledge and technological expertise and increased their professional networks. Exhibitors/businesses increased sales and networks, local communities now have practitioners with better knowledge and technologies to use in their practices, and the destinations (Sydney/New South Wales) have been able to revitalise industry sectors. Local retailers have also benefited from the business and tourist spend.

One of the most important legacies that can arise from business events is the creation and building of social capital. Social capital benefits can apply to individuals, communities and industry groups.

*Individuals* can build social capital directly by participating in events and the transformative social networks that are active and accessible for their duration. Delegates rely on meeting old contacts and making new ones through networking, formal presentations, and other activities during and immediately after the event. Time and space for this to happen may be scheduled during the event, or it can happen sometimes as a result of special initiatives, such as during research activities carried out at the event itself (Lilley, personal communication, October 2016). Seventy-six per cent of respondents agreed their research and business collaborations were supported by the conference experience; however, some sectors appeared to require a longer time to build trust than others. Other factors, such as conflicting individual research timelines, may also impact the type and strength of connections formed (Pocock, personal communication, October 2016).

The 3rd Asia Pacific Regional International Solar Energy Society Conference (ISES), for example, was held in Sydney in 2008. The conference facilitated networking and collaboration opportunities through lectures and breakout discussions where people were encouraged to expand on particular ideas. The organiser stated:

> *We specifically focused on structuring the conference so that people could discuss and build on their ideas. 80% is networking and 20% is the paper* (cited in Foley et al., 2010: 16).

*Communities* can also build social capital (directly and indirectly) through the transformative social networks that mobilise interactions and collaborations at conferences. These interactions and collaborations produce the benefits of innovation, economic development and societal change. Business events have a role in developing strong communities as they prepare for positive changes and build resilience against negative changes. Conferences can drive societal change by allowing a focus on sharing knowledge and agenda setting for medium to long-term strategic advocacy – as in the case of the IUCN Parks Congress, which only meets every 10 years. Its most recent agenda (set at the Sydney congress) was consequently titled '*The Promise of Sydney*'.

*Industry sectors*_that hold international business events can benefit from a range of opportunities which 'break the ice' and set up research collaborations with new synergies. Progress can be made on even the most complex research problems, when multiple and varied inputs are possible. During the 2010 study, delegates indicated they gained new ideas and information to take back to improve practices in their own countries, as well as an increased appreciation for belonging to a global community of people working towards the same goals. One delegate stated, "I don't feel so alone now in what I do" (Foley *et al.*, 2010: 23). In the 2015 study, 91% of respondents agreed that their event exposed them to new ideas and 93%. to new knowledge. Participants over 40 were less likely to agree their knowledge and capabilities had been developed as a result of a business event, while younger delegates (particularly those between 25 and 29) were more likely to agree.

Despite this, many senior delegates find significant value in attending international business events. Professor Irina Verenikina noted Australian delegates of all ages found the conference she organised to be useful in creating discourse between the two largest regional groups in the International Society for Cultural and Activity Research (Eastern Europe and Australasia). She said situating the event in Sydney (2014) created both intra-national and international linkages as key legacies: "It was really important that it impacted on our community to unite us more and ... also those people who came from overseas who were famous [in this area] talked to some [local] researchers" (Verenikina, personal communication, October 2016). This led to establishing new research collaborations and knowledge sharing.

Robin Aitken, New South Wales National Parks and Wildlife Service, said her experience of the 2014 IUCN Parks Congress challenged her notions of practice by being given examples of different park management practices from other countries, particularly Russia (personal communication, October 2016). Like Verenikina, she felt attending the event and being in close contact with like-minded peers gave her renewed enthusiasm for her professional practice.

The effect was also evident in the technology sector. Professor Susan Scott noted that the 2007 international conference on general relativity she hosted had a significant impact on the field, particularly by boosting researchers' morale and work outputs during a challenging time.

> *[The conference was held during] the long period while we were trying to actually achieve first detection (of cosmic gravitational waves). It was like, halfway through a 20-year*

*period. So things were getting pretty glum - a little bit, you know, we knew we were still deep in it. It was hard to get funding, it was hard to get interest, because you know there was a bit of a disbelief, like we weren't ever going to do it. Those conferences really helped keep the fire going in the middle of that two-decade period. Because we interested government, we interested the community, we interested students, and industry, and all of those things kept that spark going for the following 10 years. So it was a very pivotal part of what we were doing here in Australia.* (Scott, cited in BESydney, 2014a)

2

## Attitudinal changes

Whilst showcasing new ideas and research is an important element of many business events, it is usually not the end goal. Many stakeholders anticipating long-term legacies will go beyond the immediate 'wow' factor of local discoveries. The Association for Computing Machinery's Special Interest Group on Knowledge Discovery and Data (KDD) Mining is a premier annual conference that brings together researchers and practitioners from data mining, knowledge discovery, big data and data analytics. Organisers and speakers include representatives from Microsoft, Visa, LinkedIn, Google, Facebook, Amazon and academics from Stanford, Cornell and MIT. KDD 2015 was the first Australian edition of the event; only the second time the event had been held in the Asia Pacific region (Association for Computing Machinery, 2016). The Minister for Trade, Tourism and Major Events, the Hon. Stuart Ayres MP, said it was expected the event impact would extend well beyond revealing local achievements. He noted hosting the conference in Sydney provided an opportunity to showcase Australia's strong capability in the sector and offered local talent the ideal environment to develop new channels of communication with international operators.

The survey data revealed most delegates (83%) felt the event they attended enabled the local sector to showcase their expertise to an international audience and that it raised the international profile of Sydney. This was particularly evident in responses from European delegates. Dr Hal Lee (Polarised Neutrons in Condensed Matter Investigations Conference organiser, 2014) felt the event brought recognition of Australia as a leading research node in this field. In his interview, Lee described a series of follow-up invitations, return visits and closer collaborations, which have led to local researchers gradually being held in higher esteem by the international physics community. Lee said there were two significant and immediate outcomes resulting from the conference. The first was the increased opportunities for collaboration with researchers from Japan and elsewhere in the Asia–Pacific region. The second was the opportunity to establish new connections with European colleagues, particularly those from newly opened research facilities with similar objectives. These connections were possible because of the 'space' or 'corridors' that were provided for delegates "to present their latest developments and also find out what other people are doing" (Lee, personal communication, October 2016).

According to Lee, none of this would have happened if the event had not been held in Australia because the overseas delegates would not have been able to witness local research first hand. They also would not have had the opportunity to talk to Australian experts. Following the success of the event and the positive impression attendees were left with, it is likely that future events will be held in Australia, further enhancing the research sector's local capacity and reputation (Lee, personal communication, October 2016). Ten per cent of respondents from the technology sector, however, thought these types of benefits were yet to be realised, which again suggested potential outcomes from business events in some industries may need more time post-event to accumulate.

The study also found medical sector respondents were most likely to agree that the location of the event (Sydney) enhanced sector and community outcomes. Conference organiser, Professor William Ledger, stated in a recent interview that his event's success was aided by the location and affordability for both Chinese delegates and local junior colleagues. He noted that, from his point of view, the greatest legacy was "the links built with China and they (ISMAAR) keep talking about the meeting held in Sydney as the best one held so far" (Ledger, personal communication, 2016).

Sydney has also hosted a number of notable international medical events in the past, including the International AIDS Society conference in 2007. That conference was considered important in raising awareness among international delegates of the research and scientific excellence being carried out in the Australian HIV sector (Foley *et al.*, 2010: 11). In 2013 Sydney played host to the 2013 World Conference for Lung Cancer, with conference co-president Clinical Professor Michael Boyer stating afterwards:

> *...(the conference) led to great opportunities for [local delegates, and particularly younger delegates] to travel to the laboratories and clinical units of some of the people they met at the conference. It 's cemented relationships that have become strong research collaborations, and those things have driven the kind of research we do here in Sydney at Lifehouse, as well as right across Australia.* (Boyer, cited in BESydney, 2014b)

## Applying legacy outcomes

Innovation and application in many sectors can require individuals to take a long-term and strategic view of pursuing partnerships overseas well after the event has finished. It is likely that the total benefit is far more than that which comes immediately after attending the conference as people convert their knowledge and networking into new opportunities.

> *When people go home after a conference and they have had a good experience and made good memories and networks they continue discussions with other researchers they met here. Our university has licensed products in many countries, so the conference facilitated business links that strengthen commercialisation opportunities in the future.* (ANZSES interview in Foley et al., 2010).

In many cases, the issue of 'timing' of research (i.e. length of time between the business event and the research, or between the business event and the time it may take for a legacy to actualise) can have an impact on responses. This study found that for a number of questions, participants responded "still to be realised". Interviewees put forward a number of reasons for this, including:

- ☐ The time it takes to build trust;
- ☐ The need for incremental progress; and
- ☐ The pattern of funding cycles.

Some interviewees who had set up research collaborations as a result of attending a business event in 2014 are only now receiving funding for that project; hence they cannot yet see the legacy as having been achieved (Lee and Ledger, personal communications, October 2016). One example of such a delay is Rob Gilchrist, an Australian researcher into assisted reproduction. He attended the 2014 ISMAAR conference and only recently received a large private grant to work with Chinese collaborators whom he met at the conference (Ledger, personal communication, October 2016).

In fact, some legacies may take decades to be realised, and only with the benefit of hindsight can the real potential of business event activities be linked to positive change. In other words, business events can sometimes be viewed chronologically as reference points in time for a change of paradigm. An example of this is the 22nd International Congress of the Transplantation Society, which was held in Sydney in 2008. According to an attendee, it included discussion on topics previously seen to be only on the fringe of the sector.

> *Much of the scientific debate in Sydney centred on bioengineering and how to construct organs, rather than transplant them. We may well find that in a decade's time the Sydney meeting will be the turning point in this field.* (conference organiser, cited in Foley et al., 2010)

Given the long lead time in moving from scientific discussion and debate to the development of new techniques, technologies and practices in this field, it is likely that there are many other benefits yet to be realised from similar conferences.

## Contributions to the knowledge economy

The spiral of knowledge that flows out of a particular event is one that is worthwhile tracking over a longer period. No one has all the knowledge, and therefore solutions are produced by "the interactions of people each of whom possesses only partial knowledge" (Hayek, 1945: 530). Professor Susan Scott has articulated this sentiment in relation to hosting the General Relativity International Conference in 2011, stating:

> *We brought together the theory side of the community and the gravitational waves side, and we really need that interaction. We need the expertise from both of those sides to be passed between one group and the other, and it really made it start to happen. So that was a very pivotal time for us to have those conferences.*
>
> (Scott, cited in BESydney, 2014a)

Professor Michael Boyer has also observed that hosting the World Conference on Lung Cancer has shown him more about how his area of medical expertise progresses through event hosting and attendance. He has noted that:

> *It has made me all sorts of contacts that I wouldn't have otherwise made, it's resulted in me getting to travel to places and meet people, talk to people, and actually become aware of how important conferences – and the whole organisation that goes around them – are in the advancement of medicine.* (Boyer, cited in BESydney, 2014b)

Adequately measuring the level of global talent attracted to a destination as an event legacy also requires a longer perspective. This is because people may not be able to obtain a job immediately after a conference or to move straight away. It could also be something that burns away in the back of their minds until the time is right and the opportunity arises. As such, results indicate that this is an outcome, but one that is more difficult to document over the short term.

In summary, previous studies of business event legacies have mainly concentrated on tourist expenditure and business networking benefits. As such, broader legacies have rarely been considered and certainly not with reference to a broad range of sectors. There has also been little reference to timeframes for delivering such benefits, and the distinction between immediate and cumulative benefits, depending on the sector, or how an event might be considered pivotal in generating change. This case study is the result of research going back almost 10 years and has employed a range of research methods to uncover new insights over time.

Overwhelmingly, respondents in this study indicated attendance at these conferences enabled them to gain new ideas and knowledge. They also felt the events increased global research and business collaborations, enabled the local sector to showcase their expertise to a global audience and raised Sydney's international profile. Innovation arises from new ideas and knowledge, and academics and practitioners agree that widespread collaboration is the key to creating the innovative processes and technologies needed to compete globally and grow sector knowledge.

**Photo 2.1:** Swift Innotribe Conference 2018: Courtesy of SWIFT SC. SWIFT © 2018. All rights reserved https://www.flickr.com/photos/sibospics/44646601595/in/photolist-2b2gvc2-2b2

## Theoretical discussion

One of the crucial pillars of a thriving economy is the generation, sharing and application of knowledge. As the case study above suggests, business events provide stakeholders with unique opportunities to contribute to and participate in knowledge economies beyond the formal tertiary education environment. Conferences, in particular, are often perceived as "scientifically credible landmarks, anchors, or summaries of current knowledge" (Ferguson, 1993), which can focus attention on sometimes complex societal problems or industry-relevant issues and act as impetus for innovation and change.

However, to date, there have been few processes or tools developed to enable business event stakeholders to identify and measure such contributions or to fully understand the conditions required to ensure those contributions occur and are leveraged or maximised. Nor is there broad understanding of the range and extent of the *longer-term impacts* that can eventuate as a result of those contributions.

Where once communities dealt with localised or regional issues, our world is increasingly tackling global problems, some of which pose significant threats to our future. As we deal with these problems, the potential and quality of knowledge economies becomes increasingly important in the process of generating solutions. As such, the potential for business events to be the platform upon which economies are built also increases, and it is vitally important that we develop a more complete understanding of:

- The full extent of the potential contributions that can be made;
- The tools and resources which can best identify, measure and monitor those contributions;
- The tools and strategies that can be used by stakeholders to maximise potential contributions; and
- The longer term (long tail) outcomes that can be achieved as a result of those contributions.

In so doing, we will better equip the business event industry to plan for, promote, manage, maximise and evaluate the potential legacy outcomes that can be achieved as a result of their efforts.

## Potential benefits of identifying and maximising legacy outcomes

It is worth considering business event legacies from the perspectives of the different stakeholders who can benefit. Outcomes can vary significantly for governments and industries through to destinations and communities.

## *For publicly funded bodies and governments*

Business events have the potential to represent significantly greater value when managed and leveraged correctly and to produce a wide range of benefits. Similarities can be drawn with the approach taken by governments in investing in large infrastructure projects. When the public sector invests money into a region's infrastructure (roads, transport networks, telecommunication services etc.) they do so for two reasons. First, they see short-term economic benefit to be gained by creating local employment and business activity as projects are implemented. Second, and most important , they believe it will create long-term (future) economic and non-economic benefits as a result of, for example, increased business activity, reduced transportation costs, decreased road congestion, higher employment, increased tourism and visitation, and enhanced livability for residents, leading to population growth. In other words, there is an understanding that, whilst there may be some (relatively small) immediate economic gain (or value return) at the time of expenditure, the real economic benefit (value) will be realised over the longer term.

In the same way, whilst investment in business events may result in some short-term economic stimulus by way of increased local tourism (the tourism contribution), it is the long-term legacies and benefits (investment in local industry, knowledge expansion and sharing, connections and networks, innovative ideas and new thinking, upskilling of local economies, support for local businesses, global relationships and trade opportunities, etc.) which should be the focus when evaluating the real 'value' of an event.

Governments may therefore benefit significantly from an expanded understanding of the legacy potential of the business events in which they invest. Findings from this case study indicate that business events held in Sydney, Australia, have contributed to a wide range of positive outcomes, including progression of Australia's National Innovation Science Agenda, showcasing the local research community and facilitating collaborations and networks, and raising the international profile of Sydney. If governments (local, state and national) are able to transition from valuing business events for their visitor expenditure outputs only to leveraging them as transformative social network spaces that catalyse innovation and support long-term economic growth, they will be able to demonstrate greater value return on investment.

Many governments in developed economies are well aware of the importance of innovation in key industry sectors. Policies and programs similar to Australia's National Innovation Science Agenda can be found in almost every jurisdiction around the world, at local, regional and national levels, and there is broad acceptance that inter-organisational knowledge sharing is crucial to innovation (Kline & Rosenberg, 1986; Freeman & Liedtka, 1991). However, creating, coordinating and maintaining such knowledge sharing channels and connections locally or internationally is complex (Gereffi *et al.*, 2005).

As was demonstrated in the case study, business events – as meeting platforms – play a crucial role in facilitating these inter-organisational interactions. They are an essential vector of knowledge dissemination, knowledge creation and innovation, and their legacies extend far beyond the events themselves. Conferences have what Edwards *et al.* (2017) coined as 'long tail' outcomes. In other words, their long-term legacies significantly outpace their short-term benefits. Indeed, it is the immediate positive externalities of conferences that lay the groundwork for sustained benefits. Therefore, it is important to measure the legacies of conferences in ways which takes into account both tangible and less tangible outcomes over an extensive period of time.



## *For industries and disciplinary communities*

Business events gather the experts and professionals of a community in an interactive setting to exchange knowledge over a short period of time. These 'temporal clusters' (Bathelt & Schudt, 2010: 1958) create an information-rich ecology or 'global buzz' (Maskell *et al.*, 2004), where participants obtain and exchange up-to-date information, develop critical ideas and discover new business opportunities relative to their field just by being there and interacting with other delegates. In that context, complex messages can be exchanged between professionals, opening the possibility of obtaining immediate feedback and creating the opportunity for constructive discussions (Storper & Venables, 2004).

Participating in this 'global buzz' generates a range of positive outcomes which can be categorised into six major themes (Foley *et al.*, 2010; Foley *et al.*, 2011; Foley *et al.*, 2013). They are:

- ☐ *Knowledge gain*. This refers to the dissemination of high-quality information regarding new techniques, technologies or market trends. In conference contexts, information is easily accessible as it can be accessed through a multiplicity of channels (observation, conversation, seminars, etc.). Further, the reliability of the information is increased by the participation of multiple experts in the field (Batheld & Schudt, 2010). This knowledge can either be disseminated further in educational contexts or used in a business context to modify professional practice and generate innovation.
- ☐ *Networking*. This relates to the importance of bringing people together, establishing contacts that might turn into long-term personal relationships or strengthening/reaffirming existing contacts. Trust formed through the networking opportunities presented by conferences is central to building professional friendships and communities of practice.
- ☐ *Collaborations*. The networks developed during conferences often lead to professional and research collaborations as well as business partnerships, or transactional pipelines that can lead to increased business activity, partnerships and fruitful knowledge exchange.

- *Increased visibility*. Conferences tend to raise awareness regarding specific issues that may be relevant to a specific sector or which may highlight broader societal issues. As a result of the increased visibility, there can be a range of positive legacy outcomes, including increased funding or other support, greater community awareness and education, and increased open discussion.
- *Access to funding or capacity*. Through attracting media and political attention to global issues, conferences may help to interest corporate and political benefactors as well as investors in supporting research and other initiatives.
- *Showcase and reputation*. Conferences may be the occasion to introduce talent, recognise achievements, motivate stakeholders or participants, and increase one's reputation.

These outcomes may occur at different times (during the conference, just after the conference, or even many years later). They may be delivered in different forms and they can be evaluated from the perspective of individual delegates and/or organisations, or at an aggregate level for industry sectors and/or communities of scholars.

## For destinations and local communities

When a business event is held in a destination, it provides a relatively good opportunity for the scholars and professionals who live there to access the event, thus bringing benefits to the local community by way of increased expertise, improved techniques and new technologies. For example, when the 7th International Orthodontics Congress was held in Sydney in 2010, it attracted hundreds of local practitioners who may not have had the time and resources to attend if it had been held in another destination. The congress incorporated workshops, aimed at assisting delegates to improve their professional skills and use of new technical equipment.

Local practitioners picked up knowledge about new techniques and technologies from industry representatives who were able to demonstrate and provide information on innovative products (Foley *et al.*, 2010). The organiser noted that practitioners "place orders at the congress for new materials and implement new techniques into their practice within a month" (Foley *et al.*, 2010: 21). In a further statement the organiser noted that 'communities benefited from the advancement of the science and art of orthodontics. The majority of practitioners acquire their knowledge base from meetings. Journal publications take in excess of two years to be published and many do not read everything published in journals anyway. Knowledge is disseminated at the congress very effectively, it is filtered and presented in clinically appropriate bites' (cited in Foley *et al.*, 2010: 21).

Within a month of the 7th International Orthodontics Congress being held in Sydney, local communities were benefiting from the improved knowledge, technologies and techniques that had been brought back to workplaces by the many local orthodontists who were able to attend the conference. This phenomenon is being replicated around the world. Indeed, some forward thinking and socially responsible professional associations actively choose their conference location based upon need in the destination community for upskilled local professionals. Further, the proliferation of hybrid and virtual events post-pandemic could provide this form of community benefit to multiple locations each time a conference is held. Practitioners from around the world would then have greater access to international conferences in their fields.



## The virtuous dynamic of conference legacies

Edwards et al. (2016) proposed a framework (Figure 2.3) to aid analysis of business event legacies over time.

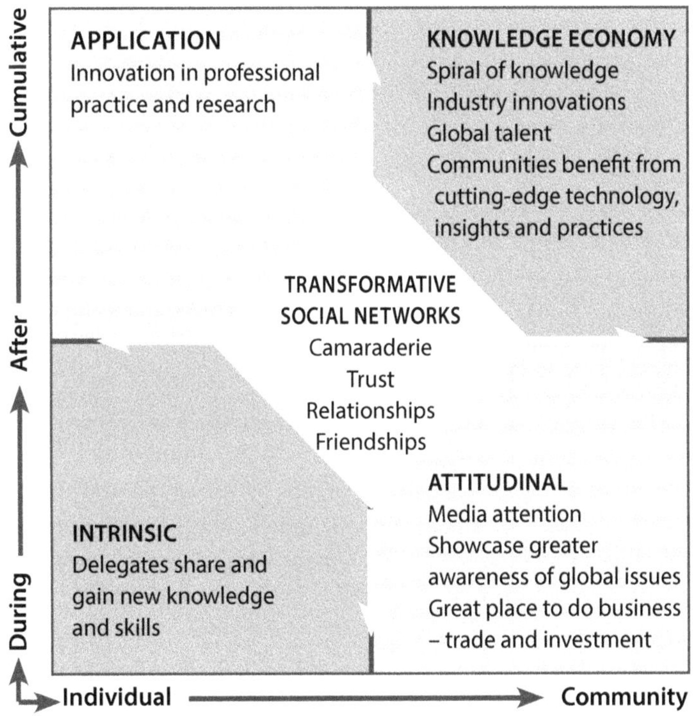


**Figure 2.3:** Virtuous dynamic of conference legacy

Known as the virtuous dynamic of conference legacy, it identifies outcomes over time using five dimensions:

1. The *intrinsic* dimension: the immediate benefits obtained by delegates or organisations during the conference;
2. The a*pplication* dimension: the long-term benefits gained by individuals and organisations who participated in the conference;
3. The a*ttitudinal* dimension: the immediate benefit to the community gathered at a conference;

4. The *knowledge economy* dimension: the long-run benefits obtained by a community and its impact on the economy at large; and
5. The *transformative social networks* dimension: the short and long-term benefits in terms of networks established by those who participated, through to the conference itself and the community at large.

Each of these dimensions is explained below.

## The intrinsic dimension of legacy

First, the primary benefit of a conference – the dissemination of existing knowledge – can immediately be experienced by participants. When they attend a conference, participants are immersed into an interactive environment where they can share their existing knowledge and gain new knowledge by listening to, giving or assisting in a presentation, conducting demonstrations or simply by exchanging ideas and knowledge with other participants informally. They can also obtain up-to-date information regarding market trends or new technologies just by being present.

They also have opportunities to meet existing or potential business partners face-to-face. Exchanges during conferences are usually fruitful as most participants tend to share a similar culture or a common knowledge base, which facilitates interaction, understanding and possibly knowledge conversion (Nooteboom, 2000). In addition, participants are escaping their daily routine, which makes them more likely to critically consider their business practices and to be open or receptive to new ideas (Borghini *et al.*, 2006).

The supportive and interactive environment of conferences facilitates the transfer of both tacit and codified knowledge (Bathelt *et al.*, 2004). Tacit knowledge is usually garnered from personal experiences and understood within a certain context. It is often implied, rather than written down in manuals, and subsequently is more difficult to share with others, particularly with those we are not in personal contact with. It is often influenced by what a person already knows (consciously or not) as a basis to gather and generate new knowledge. Codified (also known as explicit) knowledge, on the other hand, is knowledge that is transmittable in formal, systematic ways (Nonaka, 1994: 16), such as in training manuals or policy documents. The extent of knowledge creation depends on the nature of the knowledge being transferred and the process through which it is transferred (Nonaka, 1994: 18). Nonaka (1994) identified four forms of knowledge transfer:

- The transfer of tacit knowledge from one person into tacit knowledge in another;
- The transfer of tacit knowledge into explicit knowledge;
- The transfer of explicit knowledge from one source to create explicit knowledge in another source;

- The transfer of explicit knowledge from one source to create tacit knowledge in another person.

Nonaka and Takeuchi (1995: 19) developed a model to conceptualise these four knowledge-transfer processes, which indicated that the transfers take place via the processes of:



- *Socialisation* (the interactive sharing of experience) for tacit to tacit transfer;
- *Externalisation* (where hidden knowledge is progressively explicated via dialogue) tacit to explicit;
- A *combination* (where two individuals integrate their knowledge to produce new knowledge) for explicit to explicit transfer; and
- *Internalisation* (learning by doing) where a source of explicit knowledge is transferred into tacit knowledge.

If managed correctly, the conference environment provides opportunities for all four of these types of knowledge transfer mechanisms to occur, by way of workshops and practical demonstrations (socialisation and externalisation), keynotes and presentations (combination) and even simple exchanges and conversations during breaks (internalisation). In fact, it is often as a result of simple, informal conversations or exchanges between professionals in the context of business events that significant knowledge transfer occurs and is able to be used for a range of important purposes post-event. Examples of this type of knowledge transfer might include such things as research collaboration possibilities, sharing information about new funding opportunities, discussions around career opportunities, debate about proposed policy changes or simply hearing accounts of the relevant experiences of others whilst socialising.

These types of knowledge transfer opportunities enable groups and individuals to establish and strengthen contact with potential partners, and gain information, obtain instant feedback and discuss concepts with other participants. Such exchanges are also useful for measuring the compatibility of potential business relations and for creating possibilities for further exchanges. In the immediate period after the conference, participants will have gained new ideas and insights, learnt new skills, and possibly found some solutions to existing problems with potential new partners. Recent acquisition of new knowledge post-conference tends to create momentum, which often brings about an environment supportive of change and innovation (Edwards *et al.*, 2017).

## The application dimension of legacy

Over the term of their career, business event participants apply both the explicit and tacit knowledge they acquired during events to their professional practice, their businesses, their teachings and/or their research. They also transform newly established contacts into collaborations and use knowledge assets to trigger innovation. However, it is important to note that simply being exposed to knowledge or disseminating knowledge does not necessarily translate into knowledge utilisation. Knowledge acquired by those attending business events is internalised and externalised through the process of doing, and use of that knowledge can be dependent upon the individual's absorptive capacity, or their capacity to convert external information and apply it to generate innovation (Cohen & Levinthal, 1990). Hence, the *benefits* of the information acquired at a business event (for example, the range or quality of the innovative ideas that may result from it) may vary from one individual to another.

Farkas et al. (2013) presented a framework to illustrate four strategies for maximising the acquisition of new knowledge and the capacity to apply that new knowledge to actual situations. Those strategies are:

- Exposure
- Experience
- Expertise
- Embedding.

Exposure strategies relate to methods that aim to increase access to and dissemination of new knowledge, such as being exposed to the new knowledge disseminated at a business event. Experience strategies are centred on developing positive attitudes towards knowledge advancement, such as being willing to use a new or innovative strategy. The networks and contacts created at conferences can help to develop such attitudes, particularly where one of those contacts may be applying a particular element of new knowledge successfully. Expertise strategies include actions that enable one to apply new knowledge to one's professional practice (or real life), where this new knowledge is further enhanced by formal training or capacity building. Presenting at business events can be used to establish such expertise. Finally, embedding strategies are where knowledge is used to increase the use of new findings or stimulate innovation over time (Farkas *et al.*, 2003).

Applying knowledge might also include transforming contacts made at business events into actual research collaborations. Conferences are very useful channels for providing that first contact or common experience, which can help to establish what Owen-Smith and Powell (2002) call a pipeline. Pipelines are communication channels established voluntarily between spatially distant entities for exchange of information and creation of knowledge. Their formation requires investment in time and effort in order to establish a trust relationship between the participants. The creation of a pipeline for knowledge transfer tends to be more effective if the actors have previously established social contact or if they have shared a common work experience (Amin & Cohendet, 2004; Bathelt & Glucker, 2011). Being in close physical proximity at business events enables creation of these pipelines. The social aspects of conferences can also be incredibly useful in establishing the types of bonds and connections required, and if nurtured, such pipelines frequently result in important research, funding and innovative commercial projects.

## The attitudinal dimension of legacy

The attitudinal legacy refers to the positive benefits perceptible at the community level during the event. It includes the attitudes and perceptions of the participants to being a part of a large ensemble and its visibility to a larger public.

Conference attendees usually belong to a specific community of specialists who share a common interest. They are either part of a community of practice (Brown & Duguid, 1991) or are part of what Knorr Cetina (1999) calls an epistemic community. They usually share a code of conduct and similar business interests, practices and networks. This environment favours the emergence of a sense of togetherness (Bathelt & Henn, 2014). This sense of community is often capable of stimulating problem-solving, facilitating development of best practices for the field, and inspiring the strategic development of new initiatives to enhance, protect or expand the future of the community (Bathelt & Schuldt, 2010).



Through conferences, research and business communities can achieve higher visibility and attract media attention, which can, in turn, influence public perceptions and help raise awareness of specific issues for the sector. Such visibility may also generate new partnerships (either within the specialist community, with the broader community or with other sectors). In the immediate aftermath of a conference, high visibility may generate government or philanthropic support for a specific community, public funding, enhanced business practices, and increased trade and investment.

## The knowledge economy dimension of legacy

The knowledge economy legacy of conferences refers to the networks created, in which knowledge occupies a central place as a source of innovation for a community and for the global economy (Bell, 1973).

While most professionals obtain their early career knowledge by being part of a higher education environment, once graduated, development or enhancement of that knowledge comes mainly by way of their professional experiences (learning by doing) and by learning from, watching or being directed by those with whom they work. For many, the extent and quality of that learning can be somewhat limited, particularly where individuals are employed in organisations that are physically isolated (i.e. regional areas), small to medium firms (where there are few or no other employees with similar skills), organisations that are rigid and resistant to new ideas or change, or where they are operating in highly specialised sectors (and therefore not exposed to the broader knowledge developments in their industry).

In this sense, business events have the capacity to act as catalysts for the generation and sharing of knowledge beyond the initial phase of one's professional training and education. During presentations and keynotes, important, explicit information is shared to expand knowledge and generate ideas. When like-minded individuals gather to learn and share such information at business events, there is often an element of inspired momentum which acts to transcend boundaries and propel forward and creative thinking. As Hayek (1945: 530) stated, it is "the interactions of people; each of whom possesses only partial knowledge" during a conference that contribute to creating a virtuous spiral of knowledge (Nonaka, 1994).

International business events are able to disseminate existing knowledge within a sector and build global knowledge. Within a dynamic and innovative community, members benefit from being exposed to cutting-edge technology, insights and practices (Edwards *et al.*, 2016). Over time, a dynamic spiral of knowledge can have multiple and very large impacts on the knowledge economy by way of improvements to education, attracting new investments, stimulating funding and developing improved policies in a particular sector.

## The transformative social network dimension of legacy

The social aspects of business events underpin each of the above legacies and are often the catalysts for knowledge transfer and innovation. Conferences are an ideal setting in which to develop important professional and personal networks that can grow exponentially over time (Hickson, 2006). Fjelstul *et al.*, 2009 discuss the important role business events play in providing individuals with opportunities to interact with others and develop networks of professionally relevant contacts (finding they are a crucial motivator for attendance), while Bathelt and Schuldt (2008) found opportunities for these types of exchanges are richer at business events than in regular work environments.

Those exchanges can develop camaraderie, which in turn can form into friendship (Foley *et al.*, 2014c), especially if individuals meet repeatedly at annual events (Hickson, 2006). As described by Maskell *et al.*, 2005: 6, "through consecutive trade fairs and conventions, potential partners get to know one another better so that some level of relational trust can be established. Initial low risk interaction may be gradually intensified in a stepwise manner".

Genuine friendships developed over time without any initial expectation (Harris & Wheeler, 2005) can also be very important for development of future professional collaborations. According to Foley *et al.* (2014c: 54), "they establish mechanisms of trust, empathy and sympathy as common grounds of values, beliefs and interests". Some of the positive effects that have been identified as resulting from the establishment of wide professional relationships include an increase in organisational performance (Ingram & Zou, 2008), greater awareness of market trends (Deans *et al.*, 1996), the acquisition and transfer of substantial information that may have some strategic impact on organisations (Turnbull & Wheeler, 2009), increased efforts toward developing or improving cooperation (Ingram & Roberts, 2000), and evidence of greater efforts to collaborate (Ingram & Roberts, 2000).

The creation of strong social networks has been seen to help develop a 'sense of community' (Foley *et al.*, 2014c) among participants at business events and has the capacity to give a sector a sense of being united (even where the sector is comprised of individuals usually distanced from one another) in a dynamic ensemble. In this context, conferences ensure a high level of attendance and set the grounds for fruitful collaborations and blossoming innovations.

Since the 1990s, smaller face-to-face events tend to be conducted as virtual events online. With the onset of the COVID-19 pandemic, this tendency has increased. Virtual events can generate a temporary cluster (where participants are co-present in real-time online) or they can just make information available online, allowing participants access to content at any time. These types of online events do tend to generate a virtual buzz; however, that buzz lacks many of the important benefits and qualities of the global buzz that results from physical presence at international conferences (Hansen *et al.*, 2020). In the online context, participants tend not to dedicate as much attention and time to face-to-face events because they are still immersed in their daily routine (Deeter-Schmelz & Kennedy, 2002). Besides, the absence of direct interpersonal contact and the very limited two-way communication potential of online events for large numbers of participants reduce the capacity for such events to facilitate the transfer of feelings and experiences (Bathelt & Schuldt, 2010), or create trust. Nevertheless, these observations will need to be re-evaluated as people become more familiar with online events.



## Lessons learned

This chapter has presented a case study and a range of relevant theoretical perspectives which support the notion that the value of conferences extends far beyond the tourism spend associated with organiser and delegate attendance. Conferences facilitate collaborative relationships and enable delegates to share and translate their knowledge into practice and policy change once they return to their professional practice. International conferences can be perceived as the lifeblood of industry innovation, both locally and globally, in the sectors they represent. The interactions experienced at conferences are key to building transformative social networks founded on trust, friendship and camaraderie, and have the potential to underpin important future collaborations which transcend both geographical and disciplinary borders. Importantly, conferences provide practitioner delegates with enhanced knowledge, skills and technologies to better serve their respective communities.

Considering the different outcomes of conferences from a long-term perspective helps practitioners to understand the essential role they have to play in the future of delegates, organising bodies, professional communities and industries, and in the global community/economy as a whole. These long-term contributions greatly exceed the immediate benefits attained. Even if the full scope of their impact sometimes seems intangible and difficult to measure, conferences are key contributors to the knowledge economy and should be leveraged accordingly.

The case study discussed in this chapter is unique in that it draws upon research conducted on a large sample of business event delegates from a range of professions, industries and conference types. The findings, whilst not individually surprising, are strikingly similar across the various groups studied. This indicates that attitudes regarding the capacity for business events to provide accelerated knowledge creation, network development and long-term legacies are felt across a range of disciplines and areas of economic, cultural and social endeavour. The breadth of participant responses included in the data represent international and domestic visitors, males and females, later and early-career delegates, and participants from a wide range of industry sectors. From that perspective alone, this chapter provides a worthwhile contribution to the body of literature in relation to business event legacy creation and can be used as a catalyst for future discussion and research.

## References

Amin, A. and Cohendet, P. (2004). *Architectures of Knowledge: Firms, capabilities, and communities*. Oxford University Press.

Association for Computing Machinery (2016) *Conferences: Special Interest Group for Knowledge Discovery and Data Mining*. http://www.kdd.org/ conferences

Bathelt, H. and Glucker, J. (2011), *The Relational Economy: Geographies of Knowing and Learning*, Oxford: Oxford University Press.

Bathelt, H. and Schuldt, N. (2010). International trade fairs and global buzz, part I: Ecology of global buzz. *European Planning Studies*, **18**(12), 1957–1974.

Bathelt, H. and Schuldt, N. (2008). Between luminaires and meat grinders: International trade fairs as temporary clusters. *Regional Studies*, 42(6), 853–868.

Bell, D. (1973) *The Coming of Post-Industrial Society*. New York: Basic Books

Borghini, S., Golfetto, F. and Rinallo, D. (2006). Ongoing search among industrial buyers. *Journal of Business Research*, **59**(10-11), 1151-1159.

Business Events Council of Australia (2010) *Pre-budget submission 2011-12 to the federal government*, viewed 14 December 2016. http://www.businesseventscouncil.org.au/files/BECA_PBS_2011.pdf

Business Events Industry Strategy Group (2008) *A national business events strategy for Australia 2020: The business of events – Australia's untapped potential*, viewed 14 December 2016. http://www.businesseventscouncil.org.au/files/bes_ex_summary_nov08.pdf

Business Events Sydney (2014a) *Interview with Professor Susan Scott*, video file, viewed 9 December 2016. https://businesseventssydney.com.au/about-us/our-events/previous-events/besydney-ambassador-dinner-and-investiture-2016/

Business Events Sydney (2014b) *Interview with Dr Michael Boyer*, video file, 9 December 2016. https://businesseventssydney.com.au/about-us/our-events/previous-events/besydney- ambassador-dinner-and-investiture-2016/

Carlsen, J., Getz, D. and Soutar, G. (2001) Event evaluation research, *Event Management*, **6**, 247-257.

Christiaens, G., Abegglen, J. and Rowley, A. (2008) Nursing students' perceptions of a holistic nursing conference: empowerment, self-care, and application to practice, *Journal of Holistic Nursing*, **26** (2), 139-146.

Clickner, D., Martin, M., Newton, M. and Yablon, D. (1998) Nursing research conferences: benefit or burden, *Journal of the New York State Nurses Association*, **29** (1), 9-12.

Cohen, W. and Levinthal, D. (1990). Absorptive capacity: A new perspective on learning and innovation, *Administrative Science Quarterly*, **35**, 128-152.

Deans, M., Gill, D. S., and Apedaile, L. P. (1996). *The social networks of rural business owners in East Central Alberta* (Project Report 96–02). Edmonton, Canada: University of Alberta, Department of Rural Economy.

Deeter-Schmelz, D. R., and Kennedy, K. N. (2002). An exploratory study of the Internet as an industrial communication tool: Examining buyers' perceptions. *Industrial Marketing Management*, **31**(2), 145-154.

du Cros, H., Foley, C., Edwards, D. and Hergesell, A. (2017) *Catalysts for thriving economies*, University of Technology Sydney and Business Events Sydney. https://www.themeetingsindustry.org/web/wp-content/uploads/JMIC_Case-Study_BE-Sydney_Catalysts-for-thriving-economies-17.5.pdf

Dwyer, L., Jago, L. and Forsyth, P. (2015) Economic evaluation of special events: integrating economic impact and cost benefit analysis, *Scandinavian Journal of Hospitality and Tourism*, **16** (2), 115-129.

Dwyer, L., Mellor, R., Mistilis, N. and Mules, T. (2000) A framework for assessing 'tangible' and 'intangible' impacts of events and conventions, *Event Management*, **6**, 175-189.

Edwards, D., Foley, C. and Hergesell, A. (2016) *Conference: Catalysts for Thriving Economies*, Sydney: Business Events Sydney, viewed 14 December 2016. https://www.businesseventssydney.com.au/media/2070/catalysts-for-thriving-economies-research.pdf.

Edwards, D., Foley, C. and Malone, C. (2017) *The Power of Conferences: Stories of Serendipity, Innovation and Driving Social Change*, Sydney: UTS ePress.

Edwards, D., Foley, C. and Schlenker, K. (2011) *Beyond Tourism Benefits: Measuring the Social Legacies of Business Events*, Sydney: Business Events Sydney, viewed 14 December 2016. https://www.businesseventssydney.com.au/media/1366/beyond-tourism-benefits.pdf

Farkas, M., Jette, A., Tennstedt, S, Haley, S and Quinn, V. (2003) Knowledge dissemination and utilization in gerontology: An organizing framework, *The Gerontologist*, **43**, 47-53.

Ferguson, J. (1993) 'NIH consensus conferences: dissemination and impact', *Annals of the New York Academy of Sciences*, **703** (1), 180-199.

Fjelstul, J., Severt, K. and Breiter, D. (2009) An analysis of the motivators and inhibitors affecting association meeting attendance for Generation X and Baby Boomers. *Event Management*, **13**(1), 31–41.

Foley, C., Edwards, D., Hergesell, A. and Schlenker, K. (2014a) *Asian Incentive Events in New South Wales: Expenditure and Retail Impact*, Sydney, Australia: Business Events Sydney, viewed 14 December 2016. https://www.businesseventssydney.com.au/media/1599/asian_incentive_events_nsw.pdf

Foley, C., Edwards, D., Hergesell, A. and Schlenker, K. (2014b) *Estimating Inscope Expenditure Attributed to Business Events in New South Wales*, Australia: Business Events

2

Sydney, viewed 14 December 2016. www.businesseventssydney.com.au/media/1365/estimating-inscope-expenditure-attributed-to-business-events-held-in-nsw.pdf

Foley, C., Edwards, D. and Schlenker, K. (2014c) 'Business events and friendship: Leveraging the sociable legacies', *Event Management: An International Journal*, **18**, 53-64.

Foley, C., Schlenker, K., Edwards, D and Hayllar, B (2010) *A Scoping Study of Business Events: Beyond Tourism Benefits*, Sydney, Australia: Business Events Sydney.

Foley, C., Schlenker, K., Edwards, D. and Lewis-Smith, L. (2013) Determining business event legacies beyond the tourism spend: An Australian case study approach, *Event Management: An International Journal*, **17**, 311-322.

Freeman, R. and Liedtka, J. (1991) Corporate social responsibility: a critical approach, *Business Horizons*, **34** (4).

Gereffi, G., Humphrey, J. and Sturgeon, T. (2005). The governance of global value chains. *Review of International Political Economy*, **12**(1), 78-104.

Hansen, T., Pedersen, D. and Foley, C. (2020) Academic events: An empirically grounded typology and their academic impact, *Event Management*, **24** (4), 481-497.

Hayek, F. (1945) The use of knowledge in society, *American Economic Review*, **35**, 519-530.

Hickson, III, M. (2006). Raising the question #4 why bother attending conferences? *Communication Education*, **55**(4), 464–468.

Ingram, P. and Roberts, P. W. (2000). Friendships among competitors in the Sydney hotel industry. *American Journal of Sociology*, **106**(2), 387–423.

Ingram, P. and Zou, X. (2008). Business friendships. *Research in Organizational Behavior*, **28**, 167-184.

Jago, L. and Deery, M. (2010) *Delivering Innovation, Knowledge and Performance: The Role Of Business Events*, Melbourne, Australia: Business Events Council.

Kline, S.J. and Rosenberg, N. (1986) An overview of innovation. In: Landau, R. and Rosenberg, N., Eds., *The Positive Sum Strategy: Harnessing Technology for Economic Growth*, National Academy Press, Washington DC, 275-307

Maskell, P., Bathelt, H., and Malmberg, A. (2004). *Temporary clusters and knowledge creation: the effects of international trade fairs, conventions and other professional gatherings*, Annual Meeting of the Association of American Geographers, Philadelphia.

Mata, H., Latham T. and Ransome, Y. (2010) Benefits of professional organization membership and participation in national conferences: Considerations for students and new professionals, *Health Promotion Practice*, **11** (4), 450-453.

Nonaka, I. (1994) A dynamic theory of organizational knowledge creation, *Organization Science*, **5** (1), 14-37.

Nonaka, I. and Takeuchi, H. (1995) *The Knowledge Creating Company: How Japanese Companies Create the Dynamics of Innovation*, New York: Oxford University Press.

Nooteboom, B. (2000) *Learning and Innovation in Organizations and Economies*, Oxford: Oxford University Press.

Owen-Smith, J. and Powell, W (2002) *Knowledge networks in the Boston biotechnology community*, paper presented at the conference 'Science as an Institution and the Institutions of Science', 25-26 January.

Pickernell, D., O'Sullivan, D., Senyard, J. and Keast, R. (2007) *Social capital and network building for enterprise in rural areas: Can festivals and special events contribute?*, Proceedings of the 30th Institute for Small Business and Entrepreneurship Conference, Glasgow, Scotland: Institute for Small Business and Entrepreneurship, pp. 1-18.

Storper, M. and Venables, A. (2002) *Buzz: The economic force of the city*, paper presented at the DRUID Summer conference 'Industrial Dynamics of the New and Old Economy – Who Is Embracing Whom?', 6-8 June, Elsinore, Copenhagen.

Teulan, B. (2010) Conferences driving innovation, knowledge and investment, Sydney: Australia: Business Events Sydney.

Turnbull, S., & Wheeler, C. (2009). Business friendships: A free lunch? Or a source of competitive advantage. In International Conference on Managerial Solutions for the Global and Local markets, Nessebar, Bulgaria, September, 9-12.

Vega, R. and Connell, R. (2007) Librarians' attitudes toward conferences: A study, *College & Research Libraries*, **68** (6), 503-516.

Wood, E. (2009) An impact evaluation framework: Local government community festivals, *Event Management*, **12**, 171-185.

# 3 Collaboration, Partnerships and Sustainable Development

*The meetings industry needs to legitimize its role in society. This can no longer be handled by bed nights or other crude economic indicators. The narrative of the meetings industry must speak to a better world.*

Thomas Trøst Hansen, Ph.D., Science Attaché & Consul, Innovation Centre Denmark, Shanghai Royal Danish Consulate General

It is widely accepted that collaborating with other goods and service providers (for example, hotels, tour operators, food and beverage suppliers, local chambers of commerce) within a destination enhances the bidding capacity and likely success of a business event. It is also becoming increasingly clear that collaborative relationships formed with government and not-for-profit organisations (for example, charities, universities, professional associations) during the planning and execution of an event can deliver mutually beneficial outcomes for all. Finally, there is growing awareness that business events can reduce impacts and make positive contributions toward sustainability via:

- Reducing waste (through processes and product choice);
- Using local suppliers;
- Reducing carbon footprint (offering virtual options, offsetting emissions);
- Partnering with worthwhile social or environmental groups or causes (charities, environmental protection groups etc.).

The case study in this chapter relates to many of these concepts; in particular, detailing the ways in which event organisers can create important legacies for a second-tier city.

The case study is an integration of two reports, both of which investigated legacy outcomes from the 55th International Conference and Convention Association (ICCA) Congress in Kuching, Malaysia in 2016. The first (Mohamed Nor, 2017) was conducted on behalf of the Malaysia Convention & Exhibition Bureau soon after the conclusion of the congress. The second (Gaining Edge, 2020) was produced four years post event, to evaluate long-term legacy impacts.

The original bid to host this important business event was designed to support former Prime Minister Mahathir Mohamad's goal of seeing Malaysia recognised as a developed economy by 2020. Organisers planned from the outset that the conference, which is the peak event for international association meetings professionals across the globe, would not only stimulate the local economy in Kuching, but also showcase Malaysia to the industry. Attracting business events to the destination was considered a crucial step toward becoming a fully developed economy, and legacies around upskilling and educating Malaysian business event professionals were planned and delivered upon.



In our theoretical consideration, therefore, we first discuss the concepts of inter-organisational collaboration, stakeholder relations and sustainable practices in relation to the direct delivery of business events. Later in the chapter, however, the discussion moves to a deeper level where we consider whether the business events industry is positioned (and equipped) to make even greater contributions to the search for solutions to the important issues facing our world. Specifically, we look at the capacity for business events and the business events industry to become integral drivers of global collaboration and cross-sector partnerships on a larger scale than they currently are. We consider how facilitation of such partnerships might enable the business events industry to not only make an important contribution towards achievement of the United Nations' Sustainable Development Agenda, but to strengthen the authority, resilience and sustainability of the industry itself.

## Case study: Business event legacies for developing nations – ICCA Congress, Malaysia

| | |
|---|---|
| **Business event** | 55th International Conference and Convention Association Congress 2016 |
| **Location** | Borneo Convention Centre, Kuching |
| **Purpose** | Annual event to network, inform and share knowledge |
| **Primary legacy** | Development of Malaysia's business events industry |
| **Additional legacies** | Greater links and advocacy with government; education opportunities for the industry; social development (CSR initiatives); raised profile of Malaysia and Kuching as a business event destination |
| **Summary quote** | "Without the ICCA congress we wouldn't have a platform to initiate the government advocacy program" |
| **Lessons learned** | Collaboration can benefit the development of an industry sector. |

This case study was adapted from two sources: Mohamed Nor, (2017) and Gaining Edge (2020).

## Introduction

This case study investigates the benefits to Malaysia of hosting the 55th ICCA Congress in 2016. Successfully hosting a coveted international conference such as this – particularly one organised for meeting industry professionals – was seen as an important step toward developing Malaysia's business events industry, and in turn moving the country toward developed nation status.

Semi-structured interviews were conducted with 19 stakeholders from 12 organisations involved in delivery of the congress. Both research projects were conducted following completion of the business event and neither were components of the original congress planning. However, the approach was purposeful, with participants selected because they were 'information rich', leading to a greater understanding of why legacies were realised. The approach facilitated exploration rather than a measure of something (Edwards *et al.*, 2017).

## Background

In 1991 (then) Malaysian Prime Minister, Mahathir Mohamad launched Wawasan 2020 (also known as Vision 2020), declaring the "ultimate objective" was to create a "Malaysia that is a fully developed country by 2020 ... in our own mould" (Mohamad, 1997).

At that time the goal was seen to be bold and ambitious, so specific strategies were put in place to give the country the best possible chance of success. Of those strategies, establishing Malaysia as a viable business event destination and developing the infrastructure, services and skills needed to support a meetings and events industry was considered highly important. General tourism was already a key economic driver for Malaysia (see Table 3.1) but the business event sector had significant potential for growth. To achieve that growth, targets were set, including:

- Growing business tourism arrivals from 5% to 8% of overall tourist arrivals by year 2020;
- Targeting more international business events;
- Establishing Malaysia as a leading business tourism destination.

(Permandu, 2014)

Malaysia's decision to bid was integral to the strategy. The International Congress and Convention Centre Association (ICCA) is the peak body for international association meetings. With headquarters in Amsterdam, it has more than 1000 members in more than 90 countries, and with unrivalled data, communication channels and business development opportunities, represents specialists in the international meetings sector (ICCA, 2016).

**Table 3.1:** Gross national income (GNI) value by national key economic area (NKEA) sector, 2015

| NKEA Sector | GNI (RM billion) | % |
|---|---|---|
| Oil, gas and energy | 171.1 | 15.2 |
| Wholesale and retail | 168.6 | 15.0 |
| Tourism | 67.1 | 6.0 |
| Agriculture | 64.5 | 5.7 |
| Financial services | 62.7 | 5.6 |
| Palm oil/rubber | 61.0 | 5.4 |
| Electric and electronic | 53.0 | 4.7 |
| Communication content and infrastructure | 51.0 | 4.5 |
| Business services | 46.2 | 4.1 |
| Healthcare | 9.8 | 0.9 |
| Education | 9.0 | 0.8 |
| Other industries | 360.8 | 32.1 |
| TOTAL GNI | 1,124.7 | 100 |

*Source:* Pemandu, 2016: 28

In 2013, Malaysia was awarded hosting rights for the congress to be held in Kuching, Sarawak in 2016. Tourism professionals view the annual congress as a gold standard business event, similarly to how athletes view the Olympics. The congress created a platform for Kuching to showcase its capacity to host a world-class business event and impress the meeting industry professionals in attendance. Winning the bid also impacted the Malaysian government's perception of the value of attracting such business events and demonstrated how returns on government and private sector investment could be generated. The potential public relations benefit likely to accrue from hosting the congress was seen to be invaluable and almost certainly earned Malaysia credit as a serious industry player among its peers. Kuching was one of the first second-tier destinations in Asia to win hosting rights to this annual congress.

This case study looks at the legacy outcomes organisers believe were realised for Malaysia as a result of hosting this business event.

## Research methods

Given the exploratory nature of the study, a qualitative methodology was deemed appropriate. Nineteen semi-structured interviews were conducted with a range of stakeholders, including ICCA Malaysia members, product and service providers, and scholarship recipients from 12 organisations. These organisations were directly or indirectly involved in the planning and delivery of the congress. The interviews were conducted in English and lasted up to 30 minutes. The interviewees included high-level decision makers of the various organisations (CEO, General Manager, Director of Sales and Marketing) and middle managers (area managers and department heads). Interviews were held with ICCA Malaysia members from MyCEB, Sarawak Convention Bureau (SCB), Penang Convention & Exhibition Bureau (PCEB), Borneo Convention Centre Kuching (BCCK), Kuala Lumpur Convention Centre (KLCC), Putrajaya International Convention Centre (PICC), Sabah Tourism Board, and Yayasan Sabah Group. These organisations were the key drivers of the business events industry in Malaysia.

Some interviewees could not access Skype or be interviewed in person and had to provide responses via email. Some written responses required additional clarification or exploration, which again was obtained via email.

For the post-event report, interviews were conducted with the Sarawak host and participants at the 2016 Congress. For both reports, secondary data were collected, including news and media releases, email correspondence, observations and previous event reports. The data were then analysed and a thematic framework that related to established concepts was identified, allowing for emerging codes to be included in the results. The themes were examined for relationships and were further clustered into eight main categories: government advocacy, education, community engagement, ecological conservation, raising awareness and profiling ICCA Malaysia member engagement, advocacy for second-tier cities, and economic impact.

## Discussion of findings

The organisation of a major business event such as the ICCA congress requires the collaborative efforts of numerous stakeholders. The different activities, smaller business events and functional areas of the larger conference must work together as a whole. This type of integration needs thorough planning and teamwork to be carried out by relevant member groups, industry partners and other stakeholders.

The interview data revealed the significant levels of inter-organisational collaboration operating during this event resulted in numerous benefits being realised for the local business events industry, the city of Kuching and the broader Malaysian community.

Those benefits were:

- Government advocacy and political engagement;
- Industry education;
- Community engagement;
- Collaboration on social and ecological contributions;
- Awareness of the host destination;
- Greater engagement between ICCA Malaysia members;
- Advocacy for second-tier cities.

Each of these is discussed in the following sections.

## Government advocacy and political engagement

Improved relations with government and more accessible pathways on which to advocate on behalf of the Malaysian business events industry were noted as important outcomes by interviewees. At the 54th ICCA Congress in Buenos Aires in 2015, a collaborative relationship had developed between Malaysia's Foreign Affairs department and members of ICCA Malaysia. At the handover ceremony in 2016, Malaysia's Ambassador to Argentina demonstrated government support on behalf of the nation in welcoming the delegates to Kuching. ICCA Malaysia members then arranged engagement between key personnel of ICCA World and political leaders in Malaysia. ICCA Malaysia and ICCA World shared the same commitment to connect and communicate with key leaders and policy makers to raise awareness and gain support for the business events industry in Malaysia.

As a result of a number of meetings of government representatives, it was proposed that each ministry and government agency appoint or identify an internal officer, as a focal point, to provide support to the business events industry in Malaysia. This was a significant step forward and would allow for effective and efficient communication and collaboration with various government organisations to ensure the success and future development of the business events industry in Malaysia. It also resulted in widespread expressions of support for the sector from the government and political leaders in Malaysia.

## Educating the local business events industry

In support of Vision 2020, ICCA Malaysia had established the goal of developing future leaders for the business events industry. Several organisations worked together during the conference in support of this goal. In particular, they hosted young associates, students and executives, enabling them to attend the congress, gain new knowledge, network among global experts and get a better understanding of the industry as a whole. They collaborated on programs and offered grants to attract young professionals to the event. Kuala Lumpur Convention Centre offered a Professional Development Grant and hosted three young local executives. The SCB awarded five scholarships through its Asia Pacific Young Professionals Development Programme in collaboration with ICCA's Asia Pacific Chapter. The Malaysia Convention & Exhibition Bureau (MyCEB), together with the Malaysia Association of Convention & Exhibition Organizers and Suppliers (MACEOS) and Sarawak Convention Bureau (SCB), hosted three students from Sunway University, which had won the 2016 Bid Challenge initiative earlier in the year. The outcomes for the grant recipients are evident in the following quotes:

> *"(It helped me) get to know new people from different industries ... Learning something that I am unable to learn in class."*
>
> *"It's important for me to equip myself with the current knowledge in business event industry in terms of planning, additional supports, strategies, partnerships as well as personal development"* (Professional Development Grant Executive recipient).

These initiatives exposed the next generation of business event professionals in Malaysia to cutting-edge practices and the most important issues confronting their industry. They inspired and equipped them with global perspectives and tools for developing the business events industry in the future.

## Generating community engagement

For developing nations, particularly second-tier cities within developing nations, the opportunity to showcase local talent, services and facilities can be the catalyst needed to propel them forward. For the local community of Kuching, hosting the 55th ICCA Congress was just such an opportunity. Those responsible for planning and organising business meetings and events around the world would be attending, and the potential for Kuching to attract future business events to the city was enormous. Because of this, significant effort was directed toward ensuring the local community had the capacity to appropriately service delegates and that community support for the congress was evident. Interview responses from both the SCB and Borneo Convention Centre Kuching (BCCK) noted the importance of developing local talent. They said training and briefing sessions with local suppliers and providers in Kuching on the importance of hosting the congress had been conducted, highlighting what it would mean to the destination if it was deemed a success. Through the training and briefing sessions, local shops were made aware of the potential business opportunities and as a result they offered special rates for ICCA delegates and ensured provision of a wide range of products.

> *"To the shops along Waterfront Bazaar, we gave them pamphlets in English, BM and Chinese and explain and brief them about the ICCA congress and requested them to stock up on their products. We also worked with our local vendors to provide WiFi, conducted training to our BESarawak partners to ensure the smooth execution of the congress. We approached the taxi stands and also briefed them about the event"* (SCB).

'Paint the town green' was the main initiative taken by SCB (see Figure 3.1), and Kuching became ICCA's very own city for the duration of the Congress. The New Sarawak State Legislative Assembly Building, the city's most iconic building, was lit up in green. It was the first time such an initiative had been undertaken for a congress, and was achieved through collaboration with the city council. The famous cat statues in Kuching were also donned with ICCA's theme. In addition, the Sarawak Tourism Board granted permission to stage an exclusive mini Rainforest World Music Festival for the conference attendees.

Through collaboration with a local education institution based in Kuching (Segi College), MyCEB and SCB provided the Congress with 56 volunteers, with the mission to make the business event a success and to share knowledge freely. Being a world congress, there were delegates from all parts of the world. Both SCB and BCCK partnered and collaborated with three event organisers based in Kuching to coordinate and manage the social events, to provide unique and impactful experiences for the international delegates. This allowed the destination to showcase local talent and enhanced the capacity of the professional sector in Kuching. Each social event provided the delegates with a different theme and an enhanced experience.

**Figure 3.1:** Paint the town green. Source: Courtesy of MyCEB

The companies directly involved in providing products and services for the congress were also provided with a platform through which to showcase their capability and expertise, identify potential business partners, and raise their business profile.

## Collaborating for social and ecological legacies

With 800 delegates attending the congress in Kuching, the host committee had an opportunity to engage in corporate social responsibility (CSR) initiatives. Two important initiatives were the book-share project and the Borneo Orangutan Project.

### *Book-share project*

SCB collaborated with ICCA and the Librarians Association of Malaysia (Sarawak Chapter) to raise book donations to benefit children in rural Sarawak. They collected 358 books from delegates to be distributed to five schools and two villages (see Figure 3.2).

According to SCB, the book-share initiative will be adopted by future conferences. The activity has raised awareness of the potential legacies that are possible when initiatives such as this are undertaken to help rural communities in developing nations. Teachers from local universities in Sarawak are now better resourced to teach English and it is hoped this will eventually extend to rural communities.

> *"This is the first time we work on a book-share program with ICCA and now we have a few conferences that would like to adopt the book-share initiatives because of the awareness raised through the media. We are also working with few universities to teach English to adults from the rural areas so that they can transfer that knowledge to their community and so the legacy lives on"* (SCB).

*"Just imagine what changes the 358 books donated by ICCA members would bring to Sarawakians. The book-share initiative holds huge potential to change lives for the better"* (Sarawak State Library).

**Figure 3.2:** Book donations. Source: CSR initiatives bring change to communities (2016: 8)

### *The Borneo Orangutan Project (BOP)*

This project was the brainchild of both ICCA and local hosts. It was initially launched at the International Business and Tourism Management trade show in Barcelona in 2015, highlighting the plight of this endangered species. Through collaborations with Sarawak Forestry, the project focused on raising funds for the orangutan conservation efforts in Malaysia. The SCB also teamed up with University Malaysia Sarawak to design the official congress t-shirt to bring world attention to the plight of the orangutan. Six baby orangutans were adopted at various international business event tradeshows, eight-year old orangutan, Ganya, and new mother, Selina, were adopted by IMEX America and ICCA respectively via the Bronze package (RM20,000), and SKYCITY's New Zealand International Convention Centre took the Silver package (RM30,000). The BOP project not only attracted international attention but also raised RM70,000 through corporate adoptions.

These two initiatives benefited the local community and contributed to the conservation of the ecosystem. They also enhanced the reputation of the destination as being motivated to, and capable of, driving social and environmental change.

## Raising awareness of Malaysia as a business event destination

In 2015, SCB launched an integrated public relations and marketing campaign titled #RoadToSarawak. The campaign included marketing collateral, advertising, social media, and communication and collaboration with other ICCA Malaysia organisations, including Malaysia Convention & Exhibition Bureau (MyCEB), Kuala Lumpur Convention

Centre (KLCC) and Borneo Convention Centre Kuching (BCCK). To promote the campaign, MyCEB participated in business event-related tradeshows globally and built panel boards showcasing Malaysia to the attendees of the trade event. A panel board dedicated to the #RoadToSarawak campaign and booth activities organised by BCCK within Malaysia's pavilion at such events increased booth traffic and visitation.

*"We worked together to promote the event and MyCEB supported the branding of the event. There was a dedicated promotional panel at the Malaysian pavilion and everyone uploaded the #roadtosarawak hashtags on social media"* (MyCEB).

*"ICCA allowed us to bring our brand to its peak and so it's up to the rest of the industry to take that forward. Probably RM8 to RM10 million in terms of public relations and spreading the gospel of Sarawak and Malaysia as a destination"* (SCB).

**Figure 3.3:** The road to Sarawak. Source: Courtesy of MyCEB

For the first time, ICCA World did Facebook Live videos on their Facebook page and generated a reach of almost 126,000 impressions, with more than 2400 engagements and 324 link clicks. There were also three featured videos and one closing video. The number of followers for ICCA World Instagram increased by 5.5% during the conference, with 856 followers and 363 engagements generated through six posts. With 13,200 followers, the @ICCAWorld twitter account generated a reach of over 145,000 impressions, almost 5000 engagements and 396 link clicks. These outcomes indicate raised awareness in a global audience of Malaysia and Kuching as a business events destination.

## ICCA Malaysia member engagement

The ICCA Malaysia committee met regularly between 2014 and 2016 to discuss challenges and potential opportunities for collaborations that would contribute to the success of the congress.

> *"Since Malaysia won the event, ICCA Malaysia members have contributed to local host committee by submitting reports, providing feedbacks and recommendations on areas of improvements"* (MyCEB).

## Advocacy for second-tier cities

One of the main initiatives proposed during the bid was an ICCA member advocacy plan targeting second-tier cities. According to the bid, there are more than 250 potential second-tier cities in the Asia Pacific region, of which only 30 are ICCA members. This was used to reinforce Kuching's position and reputation as a second-tier city hosting world business events.

> *"What ICCA meant for Malaysia was to profile Kuching, Sarawak as a second-tier city and I think they leveraged it well"* (MyCEB).

## United Nations Sustainable Development Goals (SDGS)

Gaining Edge (2020) identified the various ways in which the business event contributed to the United Nations SDGs. The contributions made to SDGs 4, 15, 17 and 9 are summarised below.

- ☐ *SDG 4: Quality education.* A Members Gift of Love "BookShare" project, with books collected shared among five schools and two villages in rural Sarawak.
- ☐ *SDG 15: Life on land.* Awareness of the plight of the critically endangered orangutans and the push for environment conservation through the Borneo Orangutan Project.
- ☐ *SDG 17: Partnerships for the goals.* Introducing and engaging others – supply chains, ICCA delegates, etc. – to participate in the above-mentioned goals.
- ☐ *SDG 9: Industry innovation and infrastructure.* Business events sector advancement through high-quality creative programming and laying the foundations for the technological capabilities of Borneo Convention Centre Kuching, and opening doors for innovation through new technologies.

Legacy impact advocacy started during ICCA Congress 2016. As a result, in 2020, the Sarawak government recognised the long-term benefits of hosting conventions. Legacy impact was included in the Sarawak Economic Action Council (SEAC) plan for 10 years (2021 to 2030).

## Conclusion

This case study examined the development of inter-organisational collaborations which laid the foundation for sustainability of Malaysia's business events industry. The 55th ICCA Congress held in Kuching in November 2016 was a catalyst for the generation of new collaborations to assist the future development and sustainability of the nation's business events industry. Further, this business event contributed to the United Nations' 2030 agenda for sustainable development (United Nations, 2021). The case study offered evidence that suggests the inter-organisational collaborations, enhanced government advocacy initiatives, education and training opportunities, social sustainability initiatives, levels of community engagement and increased awareness of the industry/destination propelled Malaysia closer to achieving developed nation status and supported the second-tier cities' advocacy programs. The 55th ICCA Congress brought together an effective network of business event stakeholders to cultivate collaboration and integrate various objectives into a comprehensive strategy, which will serve as a template for similar business events in the future.

The education and training opportunities that the congress provided for developing talent, the marketing activities promoting Malaysia globally, and the socially responsible ecological and conservation initiatives undertaken by the conference organisers also contributed to the business event's overall success. Additionally, there was incremental economic impact for the country through the influx of new money, attributed to the spending of international delegates, with the incremental benefits adding value to the future development of the business events industry. This delegate spending impact, however, was not considered to be an important legacy outcome.

The findings have significant implications for organisers of similar business events in developing nations and the association bodies that support development programs aimed at benefiting their communities. Whilst Malaysia's business events industry is still in the developmental stages, it is becoming more robust and resilient, thanks largely to the hosting of this event and the collaborative efforts of some key industry players. Many of the collaborations (both individual and inter-organisational) that formed have continued to operate effectively and achieve shared long-term goals. Supportive networks of trust have been established among industry stakeholders in Malaysia and beyond. These networks, along with the industry training opportunities generated, have proven key to the development and sustainability of the industry.

In summary, the inter-organisational collaboration leveraged through the 55th ICCA Congress has brought together different business event stakeholders to form strong and lasting networks of collaboration and cooperation. These networks have gone on to create a comprehensive strategy that bodes well for the future. This case study has shown that, while events themselves may have only a short lifespan, the legacies which can be leveraged from them, if appropriately planned for and managed, can have social and economic benefits reaching far into the future.

# Theoretical discussion

The concepts of collaboration, partnership and sustainability have relevance to the business events industry on three levels: micro, meso and macro. At the *micro level* there is the role collaboration and partnerships play in the processes of bidding for and delivering an event, as partnering strategically with other providers, government entities or not-for-profit organisations in a destination can benefit the event itself. This level of collaboration and partnership might include partnering with suppliers (accommodation, venue, hospitality, etc.), a university or other research centre, or a government authority (supporting the event with cash, resources or promotion). We can consider these as partnerships or collaborations to increase the likelihood the event (or bid) itself will be successful.

At the *meso level* is discussion of partnership, collaboration and sustainability in terms of how these concepts relate to business events, and how the objectives, activities and outcomes of the event (or the way in which it was planned and delivered) can create broader impacts. This might include impacts on:

- Local businesses and residents via multiplier expenditure throughout the local economy, higher employment, upgraded services provided by local practitioners who brought new techniques, technologies and knowledge from the conference directly into their practice (Foley *et al.*, 2014a; Foley *et al.*, 2016; Edwards *et al.*, 2016; Foley *et al.*, 2019);
- Attendees (the quality of their event experience and the positive business, career or research outcomes that may result from collaborative relationships formed at the event) (Edwards *et al.*, 2011; Foley *et al.*, 2013; Foley *et al.*, 2014b; Edwards *et al.*, 2017);
- Not-for-profit partners and their mission (the case study examples of this are the support for orangutangs, and education with the children receiving books);
- Local tourism operators and hospitality providers (not direct partners of the event but who benefited from visitor spending);
- Government authorities (enhanced regional reputation, improved economic conditions, greater civic pride; hence continued voter support);
- The local environment (if sustainability measures were implemented).

Third, at the *macro level* of thinking, there is opportunity to consider the relationship between business events and collaboration, partnership formation, and sustainability from a systemic perspective. In other words, how do and can business events provide opportunities for collaboration and development of multi-sector partnerships across the board with potential to address the numerous significant social, economic, health and environmental problems facing the world today? This macro level of thinking considers the way business events are conceptualised, utilised, designed, measured and monitored in terms of their potential to generate and support global legacy impacts, and what practitioners, educators and industry leaders can do to increase that potential. Essentially, we consider whether business events have the capacity to become integral tools of cross-sector collaboration and global partnership formation, to not only help address the United Nations' Sustainable Development Agenda but to enhance the sustainability of the business events industry itself. The following theoretical discussion looks at all three levels of thinking.

## Micro level: Collaborating to deliver business events



"Collaboration can be said to take place when two or more parties join forces to achieve a shared objective, whether the parties are individuals, groups, businesses, institutions or nations" (Liburd, 2018, p. 9). When collaboration is used effectively through shared vision and open communication, and where stakeholders recognise the potential advantages of working together, it can resolve disputes, formulate solutions and generate benefits for entities with shared interests (Adongo & Kim, 2018). For example, destination management organisations, convention bureaux, conference organisers, venues, accommodation providers and airlines enter into collaborative partnerships to provide a seamless experience for business events. A successful bid and subsequent event bring benefits to all parties. Identifying from the outset the types of businesses and organisations that will contribute to the successful delivery of an event is a crucial stage of the planning process which should not be overlooked.

It is also important to consider both the short-term economic benefit that may be derived from a particular collaboration and the deeper and more lasting legacies that may be created (or damaged) as a result. For example, as an event organiser hoping to make delegate costs as attractive as possible, one might initially be excited if approached by a large, multi-national accommodation provider seeking to partner with the event by offering delegates heavily discounted rooms. But what if the hotel chain's CSR practices are poor, staff are underpaid, or it becomes obvious during negotiations that their only real motivation is ensuring smaller, locally owned hotels are excluded in a bid to drive them out of business?

Essentially, when business event organisers partner with other entities to plan and deliver an event, they have the capacity to generate greater support for the event, ensure greater service quality delivery, combine and share resources and promote the event more widely. Collaborating with local suppliers and providers can enable them to deliver more cost-effective and sustainable events, improve local economy outcomes and generate more local community, media and government support. But for collaborations to be genuinely successful, it is important to invest the time and effort needed to assess how well all parties' objectives, practices and expectations are aligned and to establish a partnership management process capable of supporting its ongoing development.

The possibility of realising such benefits is dependent on identifying the range of stakeholders who are likely or have the potential to be impacted by the event.

## Stakeholders

A significant amount of stakeholder research has been undertaken in the area of festivals and events in general, however less attention has been paid to stakeholders in the context of business events. Originally proposed for a corporate context, Clarkson (1995: 2) defined stakeholders as "those persons or interests that have a stake, something to gain or lose as a result of [a corporation's] activities". In the context of events, stakeholders have been defined as people or groups who have a stake in the event and its outcomes, including participants, organisers, sponsors and funding bodies, community representatives, and everyone impacted (Getz, 1991: 15). Reid (2011) argued that event managers should identify and review the agenda of stakeholders in order to balance their competing needs, tensions and expectations. Similarly, Robertson *et al.* (2009: 161) stress the need to "achieve the right balance between the needs of organisers and other stakeholders in events", while Richards (2007) points out that the compatibility of stakeholder interests is crucial to the success of any event.

Getz *et al.* (2006) built on previous research by examining case studies from various festivals to ascertain who event managers saw as their stakeholders and how they worked with them. They categorised stakeholder roles into those of regulators, facilitators, co-producers, suppliers, collaborators, the audience, and the impacted, adding that key stakeholders may have multiple roles which affect organisation of the event (Getz *et al.*, 2006). In their definition of these categories, they explained 'the impacted' referred to discrete groups such as the community, charities (receiving money or benefits from the festival) and special interest groups. However, special interest groups were described more as potential threats that event organisers could minimise by either actively seeking to involve them in the event or by at least being aware of the potential problems they may cause. For some of the festivals studied, outreach programs, charitable relationships and community giving (free tickets etc.) were initiatives aimed at engaging in "community relations to deal with issues arising from the festivals" (Getz *et al.*, 2006: 118).

Van Niekerk and Getz (2016) further developed this conceptualisation of festival stakeholders to identify eight universal groups of stakeholders (customers, competitors, employees, government, owners, special interest groups, suppliers and senior management) and 45 unique sub-groups from a multi-dimensional approach which challenged previous approaches and their skewed or hierarchical view of stakeholders. This grouping and analysis of stakeholders in festivals is highly inclusive, giving consideration to virtually all possible entities that may conceivably be impacted as a result of a festival. However, in leaving no potential stakeholder out of the model, it also becomes so broad that to realistically address all competing needs (particularly for smaller events) may be impractical. Indeed, a recent study in the UK found that festival and event practitioners had not found any established stakeholder model to be useful in their event management practice (Wallace & Michopoulou, 2019).

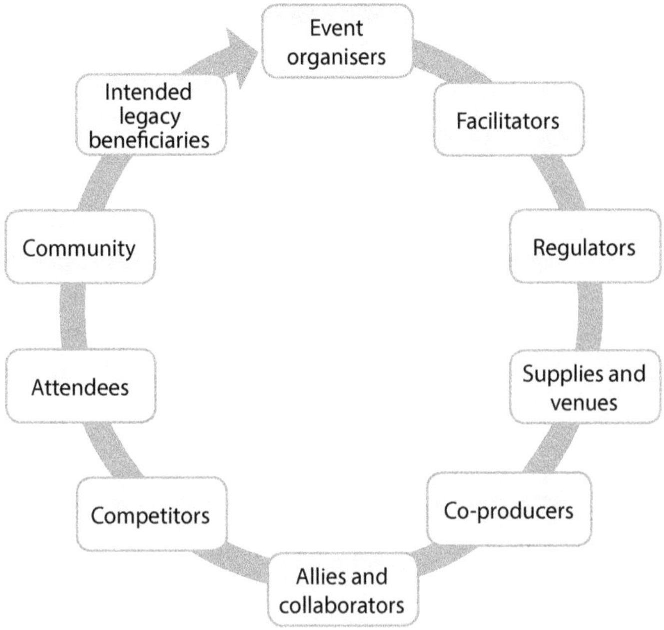


**Figure 3.5:** Business Event Stakeholder Roles. Adapted from Getz et al. (2006) and Van Niekerk and Getz (2016).

3

Such models do, however, give a comprehensive foundation from which a model of business event stakeholders could be developed. In Figure 3.5 we build on Getz *et al.*'s (2006) groupings and take elements of Van Niekerk and Getz (2016) model to propose a basis for a framework from which to categorise business event stakeholders. Importantly, this proposed framework adds a new stakeholder category, '*intended legacy beneficiaries*', which draws attention to the very important role of proactive legacy generation, which we believe business events hold. Essentially, this category refers to stakeholders one may *seek* to positively impact via strategic planning and targeted action as a result of business event legacies (i.e. the orangutans in the case study), as opposed to those who are simply impacted as a natural consequence of the event occurring (i.e. local residents who may be impacted by increased traffic). This may also include causes or issues one may proactively aim to support or benefit through event activities, such as children in areas of famine, an environmental cause, an endangered species or a quest for a global health problem. Essentially, it is going beyond thinking about the impacts an event itself is going to have (i.e. beyond thinking about how to partner with certain charities to increase the success of an event or to reduce threats to its success) and more toward thinking about the positive legacy impacts an event has the *potential* to generate for *certain stakeholders.*

Our rather simple model acknowledges that, throughout the business event lifecycle, from conception through to legacy realisation (which could be a period of many years), each stakeholder will have opportunities to wield power and influence, may be both impactor and impacted, and will have multiple relationships with other stakeholders. Whilst we provide examples for each stakeholder type below, we stress the fluidity of roles and that some entities will have multiple stakeholder roles. Each business event is unique and the roles of stakeholders are nuanced and require thoughtful management.

- **Event organisers** are those responsible for organising an event. They might include, for example, the organising body's personnel (paid staff, volunteers, interns, senior management) or owners/shareholders/board members, a higher education institution or the academics organising the event, or industry bodies and professional associations (often both domestic and international).
- **Facilitators** are the non-participating resource providers supporting the event and might include sponsors, local government agencies, convention bureaux, destination managers, legacy planners.
- **Regulators** are those determining the regulations around how an event is delivered. Generally, this group includes government authorities, law enforcement, emergency service providers, immigration, trade or industry regulators.
- **Suppliers and venues** are those who provide goods and services to the event, for example, facilities, hospitality, accommodation, promotion.
- **Co-producers** might include volunteers, other groups and individuals tangibly involved in delivery of the event, but not sponsors, suppliers, or organisers.
- **Allies and collaborators** might include, tourism agencies, associations and local businesses.
- **Competitors** might include other destinations, events or activities which may be competing for the same market segment or audience.
- **Attendees** might include delegates, industry associations, and keynote speakers.
- **The community** might include the local host community, residents, businesses, the local environment, charities, local cultural, religious, sporting or environmental groups.
- **Intended legacy beneficiaries** are those individuals or entities the event planning and delivery process is proactively seeking to impact.

This is a very broad list and it is impractical to consider all possible needs; however, it provides a starting point for consideration by business event stakeholders. The Kuching case study demonstrates how the collaboration of various stakeholders can generate legacy outcomes. We explore some of these outcomes in the following sections.

## Meso level: Sustainability practices in business event delivery

The Kuching case study includes numerous examples of legacy outcomes that business events might generate. While economic benefits derived from delegate spending have long been a primary benefit sought by those hosting large-scale business events, it was *not* considered an important legacy outcome for this event. Instead, the marketing benefits of positioning Malaysia (in general) and Kuching (in particular) as attractive and viable business event destinations, together with the education and training opportunities the congress presented for business event professionals in Malaysia, were seen to be of much greater long-term significance.



The ICCA Congress represented a valuable opportunity to raise the international profile of Kuching (a second-tier Malaysian city) and improve the likelihood of future regional growth. The case study highlights the significance of the inter-organisational collaborations established between various stakeholders throughout the planning and execution of the event, and the beneficial outcomes that resulted from those relationships. Indeed, conference organisers went to great lengths to ensure long-term value was generated, well beyond tourism expenditure. Event industry professionals were able to establish links with government decision and policy makers and there were high levels of local community engagement. They ensured important research and financial collaborations were established, and that social and ecological projects of significance to the local community (such as programs providing books for children, and charities caring for local orangutans) were supported. Finally, business event professionals from across greater Malaysia, who had previously worked somewhat in isolation, established relationships and linkages with each other during the processes of bidding for, creating and staging the event. These were significant achievements for a second-tier city.

### *First-tier and second-tier cities*

In the business events context, first-tier destinations (major business events destinations) are characterised by:

- ☐ An airport with ample nonstop national and international access;
- ☐ A convention centre with a minimum of 500,000 gross square feet of exhibit space;
- ☐ Hotel inventory, which includes major hotel brands;
- ☐ A minimum of 10,000 committable sleeping rooms;
- ☐ Abundant dining, entertainment and attraction options for attendees;
- ☐ Distinctive leisure-travel appeal.

In contrast, second-tier cities:

- ☐ Must have a regional airport with less nonstop or direct air traffic;

- ☐ Possess convention facilities with less than 500,000 gross square feet (the largest convention venue is generally a single hotel or conference centre);
- ☐ Attract a regional drive market;
- ☐ Offer greater affordability and better value.

(Destinations International, 2021).

Consequently, second-tier cities often have less experience in hosting large-scale business events and may not be as well developed in terms of inter-organisational and strategic partnerships within and beyond their destinations.

According to Fyall (2011), destinations compete more effectively by engaging with relevant industry partners to offer an integrated marketing mix and system of delivery. Destinations deal with intense competition and environmental challenges, so partnering with others is vital. Such networks are typically well developed in first-tier destinations, while second-tier cities often "lack the economic weight, political importance and the attractive pull of first-tier cities" (Cardoso & Meijers, 2016: 997). On the other hand, Parkinson (2012) noted that many second-tier cities have major concentrations of economic activity and substantial wealth creation potential, human capital and creativity. They can often provide high-quality services at more competitive price points than first-tier cities. If they have the right infrastructure, facilities, capacity and powers, second-tier cities can lift their region's economic performance and reduce inter-regional inequalities, promoting territorial and social cohesion.

This case study focuses on a business event that had the clearly stated aim of raising the profile of a second-tier city as an international meetings destination. To realise this goal, they looked to existing and new stakeholders to strengthen and build relationships. Organisers of the event were committed to this process of relationship building, and accordingly executed a successful event. Bidding for and hosting the event resulted in increased inter-organisational and international collaborations, and greater links and advocacy capacity with government.

## *Contributions to economic, social and environmental sustainability*

The 2030 Agenda for Sustainable Development was adopted by all United Nations Member States in 2015, providing a blueprint for peace and prosperity on the planet (United Nations, 2021). The vision is supported by 17 Sustainable Development Goals (SDGs) "which are an urgent call for action by all countries – developed and developing – in a global partnership" (United Nations, 2021). The vision goes on to say that a peaceful and prosperous planet will only be achievable if enough of us make contributions to the targets that will help to end poverty and deprivation, improve health and education, reduce inequality, and spur sustainable economic growth that allows us to conserve earth's oceans and forests (United Nations, 2021).

All industry areas, including the business events industry, have been called upon to make contributions to the 2016 United Nations Sustainable Development

Goals (SDGs). The case study has shown that the business events sector has much to offer in this space. The objectives of the social and ecological programs at the congress enabled Malaysia's business events industry to play a part in contributing to supporting a number of the SDGs, in particular:

- SDG 4: *Ensure inclusive and equitable quality education and promote lifelong learning opportunities for all* – by supporting quality education through the book-share project.
- SDG 9: *Build resilient infrastructure, promote inclusive and sustainable industrialization and foster innovation* – through the many industry innovations the conference brought to the destination.
- SDG 15: *Protect, restore and promote sustainable use of terrestrial ecosystems, sustainably manage forests, combat desertification, and halt and reverse land degradation and halt biodiversity loss* – through promotion and support of Malaysia's orangutan project.
- SDG 17: *Strengthen the means of implementation and revitalize the global partnership for sustainable development* – via collaborative efforts that built, strengthened and maintained relationships, and the capacities of business event stakeholders in the destination.

Jones (2014) suggests honest disclosure, transparency, communication and continual improvement are key to good sustainability management, and that even if all issues cannot be addressed at once, a staged approach to managing impacts should be implemented. The various ways in which events can contribute positively in terms of sustainable practice include operational level initiatives, such as energy use, waste reduction, transportation and water/sanitation, right through to broader issues, such as workforce labour and human rights, fair trade practices and anti-corruption, environmental stewardship and economic benefits and legacies which can be created (Jones, 2014).

Thus far, the link between business events and the SDGs has been largely ignored in academic literature. However, certain industry players have made useful contributions. In 2020 more than 40 business event organisations submitted information about their SDG contributions to the Joint Meetings Industry Council (JMIC) SDG Project. Analysis of the submissions found that SDG 12 (*Ensure sustainable consumption and production patterns*) was most commonly cited as having been achieved through initiatives such as avoiding plastic bottles, reducing paper waste, reusing exhibition stands and responsible waste disposal (JMIC, 2020). Other SDGs which were frequently referred to in the JMIC research as having been addressed by business events were:

- SDG 8: *Promote sustained, inclusive and sustainable economic growth, full and productive employment and decent work for all.*
- SDG 11: *Make cities and human settlements inclusive, safe, resilient and sustainable.*
- SDG 13: *Take urgent action to combat climate change and its impacts.*

- SDG 17: *Strengthen the means of implementation and revitalize the global partnership for sustainable development* (JMIC, 2020).

In total, therefore, business events have thus far been identified as having capacity to address 8 of the 17 SDGs at a meso level. Further growth of the business events industry's potential to contribute to this important global initiative may require action at the macro level.

## Macro level: The role of business events in stimulating partnerships to achieve sustainable development goals

Over the past 50 years, there has been a sustained shift in thinking around the benefits of inter-organisational collaboration, definitions of stakeholders and the legitimacy of businesses acting in socially responsible ways for the wider good.

The developments in technology and transport during WWI and WWII catalysed business in Western capitalist societies during the 1960s and 70s. New start-ups were everywhere, and the accumulation of economic wealth was seen as an overriding priority. Competition for resources, ideas and customers was strong, which meant corporations and organisations operated predominantly in isolation and collaborative partnerships were rare. Noted economist of the time, Milton Friedman, argued that the prime objective of managers was to generate profit and that management decisions which were justified in terms of societal good or morals were contrary to the overriding function of business (1962; 1970). During this period, the concept of 'business stakeholders' was generally limited to owners and shareholders. Theorists such as Friedman felt there were no linked interests between social good and what was good for business. In fact, he viewed maximising shareholders' profits and returns as the most important role of a manager.

By the 1990s, the concept of 'stakeholders' had been expanded to include groups such as consumers, employees, communities, governments and the environment. From then, attitudes around business ethics began to change and it became evident there were, in fact, linked interests between what businesses did and what the wider community wanted. Not-for-profits began to be seen as a type of stakeholder in how a business operated, prompting argument as to whether linking with these other types of social-good stakeholders, a business could, in turn, enhance its reputation, legitimacy and licence to operate (Freeman, 1984; Green & Peloza, 2011; Austin & Seitanidi, 2012a).

At the same time, the number of not-for-profit organisations with social justice or environmental protection agendas was growing, which meant greater competition for resources. Hence, with more businesses becoming interested in adopting corporate social responsibility (CSR) initiatives to generate business value, and more not-for-profits in need of funding, partnerships between these two previously adversarial groups began to emerge. Initially there was scepticism and mistrust from some not-for-profits, particularly those traditionally opposed to big business (Eweje *et al.*, 2020). Yet, over time, they found common ground through a process of learning, changing and working together. It also became increasingly clear that, while businesses and not-for-profits could create value individually, they were able to create *greater* value together. These collaborations were no longer a philanthropic or transactional relationship, but one of mutual contribution and benefit.

These incremental shifts were important in reconceptualising collaboration and partnerships, and laid the groundwork for the levels of collaboration needed to address the complexity and severity of our environmental and socioeconomic issues today. Individual sectors and organisations do not have the capacity to address such problems alone, and it has become increasingly clear that inter-organisational and cross-sector cooperation is crucial (Macdonald et al., 2019; Eweje *et al.*, 2020).



Some of the challenges that lie ahead for researchers are to better understand how collaborations are formed and developed, how value is created via these collaborative relationships, and how the value that is created can be tracked and measured. These challenges are of particular relevance to the business events industry, a major facilitator and driver of collaboration across the globe.

## *Cross-sector partnerships and Sustainable Development Goals*

Given this contemporary understanding and acceptance of the advantages and value of collaboration, we now consider the capacity for multi-sector, inter-organisational or cross-sector partnerships to address major environmental and social issues as identified in the UN's Sustainable Development agenda.

Of direct relevance to collaboration at the macro level is SDG 17, which seeks to '*strengthen the means of implementation and revitalize the global partnership for sustainable development*'. This goal recognises the intricacy and interconnectedness of sustainable development issues and suggests that an unprecedented level of cooperation is needed between business and non-business sectors. It has been widely influential in the promotion of the multi-stakeholder partnership approach for addressing sustainable development challenges (Macdonald *et al.*, 2019) and clearly seeks to encourage greater levels of teamwork, trust, information and resource sharing, and collaboration between individuals, organisations, sectors and countries. Indeed, the United Nations has explicitly stated multi-sector partnerships are one of the chief mechanisms for mobilising and sharing the knowledge, expertise, technologies and financial resources needed to support the attainment of the sustainable development goals in all countries (Eweje *et al.*, 2020).

Multi-sector or inter-organisational partnerships cut across sectors and may involve not-for-profit organisations, government or statutory authorities, businesses and community groups (Stokes, 2007; Zhao et al., 2012). They are typically formed when a social issue is thought to be too complex or multi-faceted for a single organisation or sector to address alone and hence there is a need to involve many stakeholders in the problem-solving process. This improves collective understanding of issues, develops more innovative and integrated solutions and ensures widespread support from important stakeholders (Macdonald *et al.*, 2019). They can be more or less formalised depending on the context and the parties or actors involved (Ada, 2013). According to Eweje et al. (2020), multi-sector partnerships involve the sharing of knowledge, consolidation of resources and the balancing of competencies of participants from multiple sectors to create synergies that result in unexpected positive and sustainable outcomes.

Interdisciplinary collaboration is essential for innovation and has a major impact on creativity because it helps to optimise the requirements of desirability, feasibility and viability; all useful in developing products, services and technologies for specific purposes (Liburd & Edwards, 2018; Moirano *et al.*, 2020). When groups or partnerships include members with diverse knowledge and backgrounds, they are able to expand their search scope, update and add new elements to their individual knowledge capital and prevent core rigidity. There is less chance they will do 'more of the same', succumb to the influence of groupthink, or fall into familiar thinking or behaviour patterns. There is also greater potential for generating innovation. As Moirano et al. (2020:5) put it, "the central argument is that the synergy of creativity is significantly enhanced by connecting competences from different disciplines".

Finally, multi-stakeholder networks and collaborations that prove effective in the implementation of SDG initiatives are based on trust relationships where partners are actively engaged in constructive dialogue, mutually beneficial learning activities and action-driven approaches (Eweje *et al.*, 2020). They also need to be supported by dynamic and purpose-driven stakeholder engagement activities that enable key participants and/or stakeholders to co-create new knowledge and develop innovative solutions to address complex sustainability issues (Eweje *et al*, 2020).

This reconfiguration of the meaning of financial value to include social and environmental values creates unique opportunities for intentional social change mechanisms to generate social and environmental impacts as forms of superior value creation (Austin & Seitanidi, 2012b). It might be argued, then, that if collaboration and partnerships are viewed as integral tools in the development of social and environmental solutions, then the processes, systems, forums or actions which foster, facilitate, create or support such partnerships (such as business events) would be equally important in terms of their capacity to create value and generate solutions. We argue that by not only inspiring and facilitating formation of partnerships, but also by tracking and measuring the ongoing impacts that result, business event professionals would be able to demonstrate more fully the significant value their activities bring.

## *Business events: Catalysts for collaboration, global partnerships and SDG achievement*

Whilst solutions to environmental and social sustainability issues are increasingly said to require multi-dimensional co-operation (Horan, 2019), Eweje *et al.*, 2020 believe the full potential of many collaborations formed thus far has not been reached. They suggest this is because the collaborations have been limited by various factors, including competitive self-interest, abortive governance arrangements, a lack of accountability, poor measurement or inadequate benchmarks for evaluation, a lack of shared goals and a deficit of trust.



Knowledge sharing and creation depend upon collaboration. Innovation, knowledge creation and learning are all best understood if seen as the result of interactive processes where people possessing different types of knowledge and competencies come together to exchange information. If the knowledge is diffuse and tacit, the more subtle forms of information exchange will only occur where people meet repeatedly in supportive environments such as meetings, conferences, conventions and congresses. However, the formation of collaborative relationships (particularly in the early stages) is sensitive to physical distance. Whilst knowledge sharing can certainly occur electronically (via telephone, video-conferencing, etc.), it is more challenging over such mediums because the trust and familiarity that comes with face-to-face interactions is missing (Edwards *et al.*, 2016).

Business events are environments in which like-minded people share information, ideas, opinions, social activities, meals and casual conversations. They are shared social experiences which take people away from their established routines. In this context, knowledge is exchanged, new knowledge and common meanings are developed, and creative ideas are generated. Business events also include intense and complex forms of interaction which involve language as well as the entire behavioural complex. This enables the establishment of common reference frames between individuals through mechanisms like rapid feedback, hand gestures, facial expressions, pointing and referring to objects in real space, subtle communication cues, information interactions and a shared local context (Edwards *et al.*, 2016). Edwards *et al.* (2016) argue that physical co-presence is important for collaboration, knowledge sharing and the formation of partnerships. They also argue that it creates the camaraderie, trust, relationships and friendships, which underpin transformative social networks, and drive collaboration, innovation and value co-creation. They add, however, that such physical co-presence need not be permanent, and that once the trust, friendship or partnership bonds have been created, they can often be maintained (to some degree at least) through other means.

As a result, it is becoming increasingly evident that business events are unique and important catalysts for collaboration and the creation of working relationships and friendships based on trust and common objectives or interests (Foley *et al.*, 2010; Foley *et al.*, 2013; Foley *et al.*, 2014b; Edwards *et al.*, 2011; Edwards *et al.*, 2016; Edwards *et al.*, 2017; Hansen *et al.*, 2020). They are integral to the formation of partnerships and by association, we believe, have a stake in the value or beneficial outcomes which result. However, until now, the development of collaborations and friendships has been largely viewed as a by-product of conference attendance, and the level of attention paid to the phenomenon by event planners may not be adequate. Foley *et al.* (2014b) note that friendships evolve where there is an audience and a setting in which to express unique aspects of one's personality and generate trust, reciprocity and a sense of belonging. Business event planners can maximise opportunities for sociable outcomes among delegates by designing warm and inviting event spaces that facilitate interaction. They can provide social spaces for the development of relationships, optimal conditions for sociability, and opportunities for play to stimulate creativity and build trust and community. International conferences are also important spaces for debating the ethical issues surrounding the nature of sustainable development and can widen understanding of how these terms can and should be used (Berchin *et al.*, 2018).

Business events, therefore, are environments which inspire, facilitate and support knowledge creation, friendships, trust relationships, collaborations, partnerships and conversations about sustainable development in a variety of contexts. The collaboration that commences before the conference begins, continues during the event and then extends beyond the timeframe and location is testament to the enduring legacy potential they hold (Edwards *et al.*, 2016). Yet, thus far, development of such relationships via business events has been largely left to chance. With a growing emphasis on collaboration and multi-sector partnerships to address the important sustainability issues confronting society today, there is significant potential for strategically designed business events to expand their role in partnership formation and become an important instrument in the development of solutions.

## New perspectives for business events practitioners and educators

Given the widespread consensus that collaboration and the formation of multi-sector partnerships are integral to addressing challenging problems, coupled with the growing body of evidence demonstrating business events are excellent environments for supporting, facilitating and generating collaboration and partnerships, it is likely that business events will become integral tools in the quest for solutions to urgent global issues.

To leverage the best outcomes possible, business events professionals, educators, students and the industry can ensure that:

- Their perspectives, approaches and activities are such that events are delivered in ways which truly maximise their collaborative potential;
- Evidence of the collaborations, partnerships, relationships and outcomes generated as a result of business events is captured, recorded and measured.

This evidence can then be used to:

- ☐ Enhance the business event industry's knowledge base (share with each other);
- ☐ Continuously improve outcomes (evaluate and evolve);
- ☐ Demonstrate the important contribution the industry is making to global social, environmental and economic issues (leverage and promote);
- ☐ Strengthen the value proposition and long-term resilience of the business events industry.



Such challenges will require implementation of a range of tools, strategies and approaches, some of which may not yet have been conceived and not all of which can be considered within the bounds of this text. One area which at least warrants discussion is the way event professionals conceptualise their role and approach to the planning of an event. Becoming a driver and catalyst in the establishment of global, cross-sector and multi-disciplinary partnerships is the most significant way in which business events can contribute to the attainment of the SDGs and hence the industry's own sustainability over the long term.

Developing robust methods of evaluating collaborative efforts may seem like a challenge. But if event organisers make explicit their objectives in terms of ongoing collaborations and partnerships, criteria to measure success and methods of evaluation will emerge.

## Lessons learned

This chapter has presented a case study and theoretical perspectives related to the ways in which strategic stakeholder collaboration can achieve substantial positive outcomes from a single, large-scale business event in the short-term, as well as how it can act as a catalyst for developing strong and lasting networks of collaboration and cooperation among key stakeholders. It provides an excellent example of how business events can benefit second-tier cities and regions, and how collaboration can benefit the development of a particular industry sector; in this case, the business events industry (local and regional).

Collaboration is vital to the processes of knowledge sharing, new knowledge creation and innovation, and is a prevalent and important feature of successful partnerships. Collaboration and multi-sector partnerships are widely regarded as integral tools in progressing the United Nations' Sustainable Development Agenda, and business events are shared social contexts which provide excellent forums for the establishment of trust-based relationships, the development of friendships, and the formation of collaborative partnerships.

Business event professionals are ideally placed to develop the collaborative capacity of business events, by cultivating new tools, processes and practices which maximise the potential for successful partnerships to arise. In doing so, the industry will be able to assert a greater value proposition and make a valuable contribution to the achievement of global sustainability initiatives.

Whilst all events (and a number of other processes) may be seen as catalysts for generating collaboration and sharing of knowledge, business events are uniquely and ideally positioned to be the drivers of multi-sector and inter-organisational partnerships capable of accommodating collaborative relationships that will generate solutions to our most pressing global social, health, economic and environmental challenges.

## References

Ada, Y. L. C. A. (2013) *How does interorganizational collaboration work? Examining interorganizational collaboration among human services nonprofits,* PhD Thesis, Illinois, US: University of Illinois at Urbana-Champaign.

Adongo, R. and Kim, S. (2018) The ties that bind: Stakeholder collaboration and networking in local festivals, *International Journal of Contemporary Hospitality Management,* **30** (6), 2458-2480.

Austin, J. and Seitanidi, M. (2012a) Collaborative value creation: A review of partnering between nonprofits and businesses: Part I. Value creation spectrum and collaboration stages, *Nonprofit and Voluntary Sector Quarterly,* **41** (5), 726-758.

Austin, J. and Seitanidi, M. (2012b) Collaborative value creation: A review of partnering between nonprofits and businesses: Part II. Value creation spectrum and collaboration stages, *Nonprofit and Voluntary Sector Quarterly,* **41** (6), 929-968.

Berchin, I., Sima, M., de Lima, M., Biesel S., dos Santos, L., Ferreira, R., de Andrade Guerra, J. and Ceci, F. (2018) The importance of international conferences on sustainable development as higher education institutions' strategies to promote sustainability: A case study in Brazil, *Journal of Cleaner Production,* **171**, 756-77.

Cardoso, R. and Meijers, E. (2016) Contrasts between first-tier and second-tier cities in Europe: A functional perspective, *European Planning Studies,* **24** (5), 996-1015.

Clarkson, M. (1995) The Toronto conference: Reflections on stakeholder theory, *Business & Society,* **33** (1), 83-131.

Destination International (2021) *Descriptors not Designators – Key attributes of 1st, 2nd, 3rd Tier Cities,* https://destinationsinternational.org/featured-articles/descriptors-not-designators.

Edwards, D., Foley, C. and Hergesell, A. (2016) *Conferences: catalysts for thriving economies,* report prepared for Business Events Sydney, University of Technology, Sydney, available at: https://www.besydney.com.au/media/1611/catalysts-for-thriving-economies-research.pdf

Edwards, D., Foley, C. and Malone, C. (2017) *The Power of Conferences: Stories of Serendipity, Innovation and Driving Social Change,* Sydney: UTS ePress.

Edwards, D., Foley, C. and Schlenker, K. (2011) *Beyond tourism benefits: Measuring the social legacies of business events,* report prepared for Business Events Sydney, University of Technology Sydney, https://www.besydney.com.au/media/1619/beyond-tourism-benefits.pdf

Eweje, G., Sajjad, A., Nath, S. and Kobayashi, K. (2020) Multi-stakeholder partnerships: a catalyst to achieve sustainable development goals, *Marketing Intelligence and Planning*, **30** (2).

Foley, C., Edwards, D. and Hergessel A. (2016) *Asian incentive events in New South Wales: Expenditure and retail impact*, report prepared for Business Events Sydney, University of Technology Sydney, https://www.besydney.com.au/media/1616/asian_incentive_events_nsw.pdf

Foley, C., Edwards, D. and Hergessel, A. (2019) *Direct expenditure attributed to international conferences held in New South Wales 2017-2018*, report prepared for Business Events Sydney, UTS Sydney, https://www.besydney.com.au/media/3360/bes002_de6_conferencesinfographic_a4-2.pdf

Foley, C., Edwards, D., Schlenker, K. and Hergessel, A. (2014a) *Estimating Inscope expenditure for business events held in New South Wales*, report prepared for Business Events Sydney, Sydney: UTS. https://www.besydney.com.au/media/1617/estimating-inscope-expenditure-attributed-to-business-events-held-in-nsw.pdf

Foley, C., Edwards, D. and Schlenker, K. (2014b) Business events and friendship: Leveraging the sociable legacies, *Event Management*, **18** (1), 53-64.

Foley, C., Schlenker, K., Edwards, D. and Hayllar, B. (2010) *A scoping study of business events: Beyond tourism benefits*, report prepared for Business Events Sydney, UTS, Sydney, http://hdl.handle.net/10453/16848

Foley, C., Schlenker, K., Edwards, D. and Lewis-Smith, L. (2013) Determining business event legacies beyond the tourism spend: An Australian case study approach, *Event Management*, **17** (3), pp. 311-322.

Freeman, R. (1984) *Strategic Management: A Stakeholder Approach*, Boston: Pitman.

Friedman, M. (1962) *Capitalism and Freedom*, Chicago: University of Chicago Press.

Friedman, M. (1970) The social responsibility of business is to increase its profits, *New York Times Magazine* (13 September), p. 32).

Fyall, A. (2011) Destination management: Challenges and opportunities, in Y. Wang and A. Pizam (eds), *Destination Marketing and Management: Theories and applications*, Wallingford, UK: CAB International, 340-357.

Gaining Edge (2020) *Legacies from ICCA Congress 2016 Kuching SARAWAK: How hosting ICCA 2016 impacted our future*, report prepared for Business Events Sarawak and Borneo Convention Centre.

Getz, D. (1991) *Festivals, Special Events and Tourism*. New York: Van Nostrand.

Getz, D., Andersson, T. and Larson, M. (2006) Festival stakeholder roles: Concepts and case studies, *Event Management*, **10** (2-3), 103-122.

Green, T. and Peloza, J. (2011) How does corporate social responsibility create value for consumers?, *Journal of Consumer Marketing*, **28** (1), 48-56.

Hansen, T, Pedersen, D. and Foley, C. (2020) Academic events: An empirically grounded typology and their academic impact, *Event Management*, **24** (4), 481-497.

Horan, D. (2019) A new approach to partnerships for SDG Transformations, *Sustainability* **11** (18), p. 4947. https://doi.org/10.3390/su11184947

ICCA (2016) Country & City Rankings – Public Abstract. International Congress and Convention Association.

JMIC (2020) JMIC + SDGs: Meeting the world's Sustainable Development Goals, available at: https://www.themeetingsindustry.org/web/wp-content/uploads/JMIC-SDG-Report-2020.pdf

Jones, M. (2014) *Sustainable Event Management: A practical guide,* London: Routledge.

Liburd, J. (2018) Understanding collaboration and sustainable tourism development, in J.J. Liburd & D. Edwards (eds), *Collaboration for Sustainable Tourism Development,* Oxford, UK: Goodfellow Publishers, pp. 8-35.

Liburd, J. and Edwards, D (2018) *Collaboration for Sustainable Tourism Development,* Oxford, UK: Goodfellow Publishers.

Macdonald, A., Clarke, A. and Huang, L. (2019) Multi-stakeholder partnerships for sustainability: Designing decision-making processes for partnership capacity, *Journal of Business Ethics,* **160** (2), 409-426.

Mohamad, M. (1997) Vision 2020: The way forward, Keynote address at the Vision 2020 National Congress, Petaling Jaya, 29 April, Prime Minister's Department, Kuala Lumpur.

Mohamed Nor, M. (2017) *The sustainability of the business events industry in Malaysia: Leveraging inter-organisational collaboration for the 55th ICCA Congress, Malaysia Convention & Exhibition Bureau.* https://www.themeetingsindustry.org/web/ wp-content/uploads/ JMIC_Case-Study-Program_MyCEB_ICCA-congress-17.5.pdf

Moirano, R., Sánchez, M. and Štěpánek, L. (2020) Creative interdisciplinary collaboration: A systematic literature review, *Thinking Skills and Creativity,* **35**, 1006-26.

Parkinson, M. (2012) *Second Tier Cities in Europe: In an age of austerity why invest beyond the capitals?* Liverpool: ESPON

Pemandu (2014) *Economic Transformation Report 2013.* Report produced for the Prime Minister's Department, Putrajaya, Malaysia.

Pemandu (2016) *Annual Report 2015.* Report produced for the Prime Minister's Department, Putrajaya, Malaysia.

Reid, S. (2011) Event stakeholder management: Developing sustainable rural event practices, *International Journal of Event and Festival Management,* **2** (1), 20-36.

Richards, G (2007) Culture and authenticity in a traditional event: The views of producers, residents, and visitors in Barcelona, *Event Management,* **11** (1-2), 33-44.

Robertson, M., Rogers, P. and Leask, A (2009) Progressing socio-cultural impact evaluation for festivals, *Journal of Policy Research in Tourism, Leisure and Events,* **1** (2), 156-169.

Stokes, R. (2007) Relationships and networks for shaping events tourism: An Australian study, *Event Management,* **10** (2-3), 145-158.

United Nations (2021) The 17 Goals, viewed 12 March 2021. https://sdgs.un.org/goals

Van Niekerk, M. and Getz, D. (2016) The identification and differentiation of festival stakeholders, *Event Management,* **20**, 419-431.

Wallace, K. and Michopoulou, E. (2019) The Stakeholder Sandwich: A new stakeholder analysis model for events and festivals, *Event Management* **23** (4-5), 541-558.

Zhao, K., Yen, J., Ngamassi, L. Maitland, C. and Tapia, A (2012) Simulating inter-organizational collaboration network: A multi-relational and event-based approach, *Simulation,* **88** (5), 617-633.

# 4 The Economic Legacies of Business Events – Measuring the Iceberg

> *We know as leaders in our industry that events do create legacies for individuals and destinations, but it's only recently that we're really starting to prove what those legacies are, and realising the truth behind our words.*
>
> Tracy Halliwell MBE, Director of Tourism, Conventions & Major Events, London & Partners

In the early stages of the business events industry there was scant directly relevant literature on ways of measuring the economic impacts generated by events. Research from other fields was used to inform economic impact studies, however many of these studies used sometimes conflicting and inaccurate assumptions about visitor expenditure. This made it difficult for practitioners to accurately determine the economic impacts generated. Over recent decades, however, the industry has become an important area of economic activity. As it has grown, so too has scholarly interest around development of tools and processes to accurately measure the economic contributions made (e.g. Dwyer *et al.*, 2000; Jago & Dwyer, 2006; Jago, 2012), and to a lesser extent, the non-economic ones (Deery *et al.*, 2005; Foley *et al.*, 2010; Foley *et al.*, 2014a).

This new research is providing deeper understandings of legacy creation and the broad range of economic advantages possible when business events are strategically planned, leveraged and delivered. Many industry players are recognising that tourism/delegate spend is just the tip of the iceberg in terms of the economic benefits their events can generate and are developing clever initiatives aimed at creating long-term economic advantages for stakeholders. Business events are being used to energise or revitalise local industry sectors, attract external entities (residents, businesses or employment talent) and enhance international reputations in industries, activities or pursuits. Whilst the impacts of such activities are not immediately apparent, it is now understood they often bring significantly higher long-term economic benefit and, hence, warrant further study.

The following comment from the Director of Tourism, Conventions & Major Events at London & Partners illustrates why it is important to understand the full economic impact of business events.

> *The events industry is constantly questioned – by marketeers who question the importance of live events as part of their marketing budgets, by economic development organisations who have to balance time and resource on promotional activity, by governments who need to decide where to concentrate their efforts and funds, and by individuals who may be deciding on their future careers. Business events not only contribute to tourism economic benefit for a destination, but they also help to underpin economic growth for countries, highlighting sectors of excellence, attracting investment, and facilitating education and networking opportunities.*
>
> (Tracy Halliwell MBE personal comment, March 2021)

**Photo:** London Tech Week, Tech Night Party entrance at Shoreditch Town Hall (courtesy of London & Partners)

The following case study examines the performance of London Tech Week in terms of creating long-term economic, commercial and destination benefits for London. Essentially, the event aims to position London as Europe's tech capital by raising awareness within the sector of the opportunities London offers around innovation, employment and investment. The case study looks at the 2017 and 2016 events and provides insights into Gross Value Added (GVA), Foreign Direct Investment (FDI) and messaging outcomes of the event. It also discusses job creation, reputation enhancement and the long-term social, economic and community benefits generated. Against the background of the case study, broader issues relating to economic impact assessment, both at national and sub-national level, are discussed later in the chapter, as well as ways to plan for, track and measure the economic contributions of business events.

## Case study: Economic legacies of London Tech Week

| | |
|---|---|
| **Business event(s)** | London Tech Week 2016 and 2017 |
| **Location(s)** | London, United Kingdom |
| **Type of business event(s)** | Europe's largest business event of live tech, presented as a week-long series of events held in various venues across London for close to 50,000 visitors |
| **Business event purpose** | To celebrate London's booming, diverse and vibrant techscene and showcase the best of the city to attract new talent |
| **Primary audience** | Direct foreign investment entities seeking to invest in London tech enterprises and international tech companies looking to expand there |
| **Primary business event legacy** | To position London as the tech capital of Europe |
| **Additional legacies (beyond tourism)** | Job creation within the tech sector. Raise awareness within the tech sector about opportunities the London tech industry offers. Create long-term economic benefits for the city. |
| **Summary quote** | "The business event has added value to London as a destination for tech talent, innovation and investment by showcasing the best of tech … and creating opportunities." |
| **Lessons learned** | More comprehensive and long-term evaluative tools are needed to appropriately plan for intended outcomes and to measure short and long-term impacts. |

This case study (CS) was adapted from: Thiebault, A. (2018).

## Introduction

London Tech Week is Europe's largest festival of live tech events. This large-scale business event showcases hundreds of unique events taking place in different venues all across the city and represents the entire local technology ecosystem. In 2016, more than 300 events brought together 40,500 people, while the 2017 festival attracted close to 50,000.

Whilst many who visit this unique and multi-layered event may view it as little more than a great way to celebrate London's booming, diverse and vibrant tech scene, those working behind the scenes have greater goals front of mind.

By bringing together the best the region has to offer in terms of technological brilliance, innovation and potential, and combining it with financial development opportunities, business strategy advisors and the magic that is London, organisers hope to use the event to bring investment, knowledge, talent, job growth and enthusiasm to London and to position the city as the focal point of technological development in Europe.

## Background

*Our world is becoming increasingly reliant upon technology and the rate of change and development in the industry is remarkable. Whether it's improved medical technologies such as artificial organs, advanced communications bringing us increasingly powerful global connectivity, or the growth of artificial intelligence and smart homes – technology is reshaping the way we do things.*

*London Tech Week (previously London Technology Week) is a celebration of all things technology, London style. The event was first held in 2014 and has quickly become a key event in the United Kingdom's (UK) technology industry profile. This business event (which actually includes a suite of hundreds of live tech events) is designed to connect the entire tech ecosystem both within London and beyond, shape the future of the industry and create business opportunities that help maintain and grow London's tech sector.*

*These objectives are in line with the Mayor of London's priorities to "protect and enhance London's global economic attributes and competitiveness". Technology is a significant part of those attributes, and the purpose of London Tech Week is to showcase that aspect of the city's strength.*

*London Tech Week 2016 took place 20-26 June 2016. It was organised and supported by:*

- London & Partners (L&P): an international trade, investment and promotion agency for London ;
- UBM: a British business-to-business events' organiser based in London (now Informa);
- Tech London Advocates: a network of more than 9000 tech leaders, entrepreneurs and experts in the UK, and over 50 countries worldwide;
- ExCeL London: an exhibition and international convention centre in East London.

Other key strategic partners were Tech City UK, techUK and UK Trade & Investment (known now as the Department of International Trade).

The face-to-face networking, discovery and learning that occur throughout the event are designed to create an inclusive and sustainable growth environment for tech organisations and social enterprises of all shapes and sizes in the region. Organisers claim no other live business event brings together so many domestic and international tech specialists and enthusiasts to take part in such a variety of networking, social, learning and business opportunities.

The unique nature of London Tech Week is that it is a crowd-sourced business event with 200–300 different discrete events under its umbrella, ranging from small networking events to conferences, a hackathon and trade shows. For the 2017 business event, the criteria for hosting the event were that it must:

- Take place in London within the M25 boundary;
- Take place 12-16 June 2017;
- Be categorised under one of the six topic streams: business, digital disruption, innovation, security, start-up and scale-up, and diversity.

**Photo:** Networking at Tech Night Party, courtesy of London & Partners

Business events could include (but were not limited to) networking drinks or dinners, breakfast briefings, workshops, hackathons or codefests, product launches or demonstrations, round tables and more. The event host decides on the duration of the event as long as it is within the time frame of London Tech Week. It can be a short one-hour networking event or a three-day conference for example. Potential business event attendees can browse the business event website or the mobile app for different events happening during the week. They can filter by event stream, dates or price (free or paid).

## Research context

The research upon which this case study is based was provided by London & Partners. London & Partners is one of the founding partners of London Tech Week. They work closely with KNect365 and Tech London Advocates to deliver the business event each year.

London & Partners is the Mayor of London's official promotional agency and supports the Mayor's priorities by promoting London internationally. They do this by devising creative ways to amplify the Mayor's messages, priorities and campaigns to international audiences. Some of their objectives include:

- Helping overseas businesses set up in London;
- Helping London businesses win contracts overseas (through the Mayor of London's International Business Programme);
- Supporting companies to grow in London through growth initiatives;
- Attracting capital investment for regeneration projects.

London & Partners promote London as the first choice for overseas entities expanding into Europe or the UK and guide businesses (from start-ups to international companies) through the relocation process. The successfulness of their activities is measured by the additional economic impact generated as a direct result of their intervention. This is defined as the extra benefits generated compared to what would have occurred had they not been involved. More specifically, those benefits are the GVA and jobs created. Measurement of GVA is done by surveying audiences. More detail on GVA calculations can be found in the research methods section.

## Research methods

This case study evaluates the 2016 and 2017 London Tech Week business events in terms of the:

- Direct expenditure (short-term economic benefit) generated for London;
- Potential direct foreign investment (long-term economic benefit) brought to London;
- Professional and commercial benefits created by raising the profile of London as a tech hub and showcasing the wealth of talent throughout the city's technological ecosystem;
- Professional development opportunities created for the local (tech) industry sector;
- Value which participants and attendees perceived the event offered.

To do this, a number of secondary data sources were utilised. They were:

- A GVA calculation to measure direct business event expenditure
- Data from London & Partners' Foreign Direct Investment (FDI) team on leads generated as a result of the business event
- Data from the London & Partners' Insights team to evaluate the development of London's tech sector, and
- Survey findings via Crimson Hexagon (a social media tool) to measure sentiment on London's tech sector within the London Tech Week social conversations.

*GVA calculation*: London & Partners' main measure of success is the additional economic impact created, or the extra benefits generated as a direct result of their interventions. They measure the GVA by surveying a representative sample of audiences. The direct economic impact is driven by the spend of business delegates and event organisers in London, which is converted into additional GVA by accounting for L &P's activities and by using Input-Output tables to convert spend into GVA.

Average gross spend varies depending on the combination of the event's characteristics, with international, multi-day events generating higher value than domestic single-day events. In order to calculate the impact of London Tech Week, therefore, the following factors were taken into consideration and captured in the business event's economic impact model:

- Number of total events hosted during the week;
- Length of each event (single-day or multi-day events);

- ☐ Size of the event (measured in number of attendees);
- ☐ Attendee composition (the country of origin of most attendees – which has an impact on their average daily spend).

Multipliers that attribute an economic impact for each delegate and attendee's spending (depending on the combination of the above factors) have been developed based on primary research with event organisers and attendees. These multipliers were used to estimate the total value of London Tech Week for London.

More detail on the way London & Partners evaluates the GVA impact of their activities on London can be found on their website.

*Foreign direct investment team data*:

This data reports on the number of leads and projects acquired by the London & Partners FDI team as a result of London Tech Week events. To be included in these figures, the FDI team has either organised or participated in events during London Tech Week and met with key contacts, some of whom become leads for new projects in London (setting up or expanding). The FDI team provides feedback on each event they participated in and each lead that was generated as a result. The information is then fed into the I&P dynamics CRM system.

Whilst the results from the 2016 business event are more comprehensive, early figures from the 2017 business event are not an estimate, but from an accurate activity report provided by the FDI team. Further details can be found in the discussion of findings.

*Insights team data*:

London & Partners' Insights team monitors foreign direct investment completions during the year and works with partners to get useful statistics on sectors of importance, such as the digital/technology industry. To estimate the size of this sector in London and its evolution over the last few years, the Insights team requested the help of Oxford Economics. Data on jobs, GVA, productivity, business investment and stocks since 2012 were prepared by Oxford Economics on a bespoke basis (figures shown in Table 4.4).

*Crimson Hexagon media data*:

One of the reasons London Tech Week was founded was to build London's reputation for tech. In order to measure how successfully the event achieves this objective, L&P measures the impact of the business event on people's sentiment about the London tech scene. Monitoring activity and comments on social media was chosen as an appropriate means of evaluating this impact. Conversations about London Tech Week and related sentiment on the London tech sector are included in the data, which was gathered and reported on by Crimson Hexagon, a social media sentiment tool.

Crimson Hexagon's platform uses proprietary, example-based statistical classification to quantify customer sentiment and topics of social conversation in near real time. It combines human judgement and computer automation to identify key themes and is capable of accuracy as high as 96% in any language.

## Findings

The bulk of the findings relate to the 2016 business event. Some of the 2017 highlights related to economic benefits to the city are included, however they are scant as London & Partners was still in the process of collecting 2017 post-show data at the time of case study preparation. Table 4.1 gives basic statistical data from the 2015, 2016 and 2017 business events.

**Table 4.1:** Key statistics from the 2015, 2016 and 2017 London Tech Week business events

| Year | Attendee number and profile | No. of events | No. of event hosts |
|---|---|---|---|
| London Tech Week 2017 | 48,170 from over 93 countries | 219 | 171 |
| London Tech Week 2016 | 40,520 from 74 countries | 302 | 210 |
| London Tech Week 2015 | 40,000 attendees | 228 | 141 |

### Benefit findings related to GVA (short-term economic benefit)

These are essentially the first-round economic impacts associated directly with the event such as tourism and services expenditure. The GVA calculated for London Tech Week 2016 was £2,531,752 and £3,615,721 for 2017. Based on this calculation, the total economic benefit for London was also obtained, using multipliers. The economic benefit captures secondary impacts, and as a minimum this will include the indirect impacts, which will capture any supply chain impacts. The secondary economic benefit calculated for London Tech Week 2016, therefore, was £21,218,425 and £45,915,245 for 2017.

**Table 4.2:** FDI leads as a result of London Tech Week 2017 (preliminary results)

| Events attended / participated in / organised | Number of leads generated | Estimated jobs created in Yr 1 by the company | Estimated jobs created in Yr 3 by the company |
|---|---|---|---|
| Networking events at Upgroup reception at the Swedish Embassy | 5 new active leads | 5 jobs total | 5 jobs total |
| National Tech Transfer East China Centre, European Talent Centre/Career Frog UK office launch | 1 new lead | n/a (lead stage) | n/a (lead stage |
| 8th Sino-European Entrepreneurs Summit at Grosvenor House | 4 new leads | 9 jobs total | 20 jobs total |
| China-UK Hi! Technology Business Event at Cocoon Network | 3 new leads | 6 jobs total | 15 jobs total |
| Guangdong Zhongshan Delegation Meeting | 3 new leads | 4 jobs total | 11 jobs total |
| Toronto Smart Cities delegation | 1 new lead | n/a (lead stage) | n/a (lead stage) |
| Scaling up in London and the UK | 2 new leads | n/a (lead stage) | n/a (lead stage) |
| Set-up clinic with commercial partners for delegations from EMEA/LATAM | 5 leads converted into projects | n/a | n/a |

## Benefit findings calculated from FDI outcomes

In addition to celebrating the London tech industry and its businesses and talent, London Tech Week fosters foreign direct investment into London. One of the key sectors that London & Partners caters to is the technology sector. London Tech Week has therefore become a vector for their FDI team to make new contacts within the sector, and to meet with existing contacts to help them set up or expand their business in London.

London Tech Week 2017 had only recently taken place at the time of preparation of this case study, and results re FDI are only preliminary. Outcomes recorded thus far (in terms of businesses wanting to set up or expand in London) were, however, relatively positive. Table 4.2 is a summary of the early results.

## Benefit findings from Crimson Hexagon data

Sentiment analysis was conducted on conversations about London Tech Week 2017 on social media. The time-period looked at was from 1 November 2016 (when the 2017 edition of the business event was launched) to the end of June 2017 (when London Tech Week ended). The analysis included monitoring of blogs, forums, news sites and the following social media channels: Twitter, Facebook, YouTube, Instagram and Tumblr. Twitter was the main data focus, accounting for 92% of the overall online conversations. To single out the social media posts related to the London Tech Week conversations, the following keywords were entered into the system: "london tech week", "#LondonTechWeek", "#LTW", "Londontechweek.com", "LDNTechWeek", "@LDNTechWeek", "LTW17", "LTW2017", "london technology week" and "#LTWScaling". In order to get the most relevant results, we excluded some keywords such as "#LTWTattoo" and "@LTWTatoo" which were used by other social accounts for different activities. We also excluded the Arabic and German languages where conversations using some of these hashtags were used for other purposes. The total number of posts that resulted from that search was 81,841.

Using the system's algorithm, all social posts were then isolated around the London tech scene/sector by categorising a large sample of relevant posts. The total number of relevant posts upon which the sentiment analysis was based was 40,267.

By identifying samples of negative, neutral and positive posts, the proportion of posts sending a positive message about the London tech scene was able to be measured. The bulk of these posts included comments which demonstrated the business event fostered enthusiasm and shone a light on the vibrancy and growth of the London tech sector. Positive messages about the London tech scene were isolated and included comments such as: "a thriving technology sector" or "I love working with London Tech Scene." Expressions where there was support for the industry, pride in the growth of the sector, appreciation for the quality of the people in the industry, or the benefits the sector brought to the economy, etc. were the types of comments included in this analysis. Expressions of surprise were not taken into account. Results of social media sentiment are shared in the next section.

## *London Tech Week 2016 results*

Because London Tech Week hosted over 300 tech events in 2016, there were many opportunities for the London & Partners FDI team to connect with businesses from the sector. In most cases, the FDI team members contacted potential leads before the business event and arranged meetings at particular events. For events where the team was directly involved (organising, supporting or arranging it with partners etc.), they accessed an event attendee list beforehand. By having a list of companies that planned to attend the event, contacts could be made directly with those companies (or through the partners) before the event. After making contact, they could explain what they offered for businesses interested in expanding into, or establishing themselves in, London.

Table 4.3 shows the projects the FDI team reported supporting and helping as a result of London Tech Week 2016. These projects are now recorded in London & Partners' Dynamics CRM system and will be regularly followed up on until completion. When a project reaches the completion stage, it means London & Partners has received a completed feedback form from the company, they have set up their company with Companies House and they are a legal entity operating in the UK with a physical address.

**Table 4.3:** FDI projects created/completed as a result of London Tech Week 2016

| Project/Business | Estimated jobs created in Yr 1 by the company | Estimated jobs created in Yr 3 by the company |
|---|---|---|
| Finnish company | 1 | 1 |
| Chinese company | 3 | 10 |
| Chinese company | n/a | n/a |
| Brazilian company | n/a | n/a |
| South African company | 2 | 5 |
| Chinese company | 70 | 100 |
| Chinese company | 70 | 150 |
| US company | 5 | 15 |
| Colombian company | 2 | 3 |
| Israeli company | 10 | 10 |

They also complete a feedback form which details how many jobs they expect the new operation or expansion will create. Table 4.2 shows that London Tech Week helped 10 new companies set up in London and contributed to the London and UK economy. It also shows that new jobs will be created in London as a result, and that new tech talent is being driven to the city. Although the increase in the number of jobs, GVA and investment in London shown above cannot be directly attributed to the business event, findings from this research suggest London Tech Week does generate interest and investment in the sector.

## *London Tech Week 2017 results*

According to organisers, the 2017 event was "bigger and bolder" than previous events. It had a mission statement which read:

> *"Placed in the heart of the tech capital of Europe, London Tech Week brings together over 55,000 of the best creative and tech minds for an unrivalled mega-tech (business event). With a community platform hosting leading-edge digital content, learning opportunities and engagement throughout the year, London Tech Week delivers a vehicle to create a lasting legacy for tech talent and business"*
> (London Tech Week, 2017).

The objectives of the event were to showcase that #LondonIsOpen in a post-Brexit world would drive business and attract talent, cultivate London as a global powerhouse of tech and set the pace for technological innovation. Findings from post-event data indicate the event did in fact bring economic, business and professional benefit to London by attracting direct foreign investment into the city as tech companies set up in London or expanded as a result of participating in the event.

## Key benefits realised

The figures in Table 4.4 show the positive evolution of the tech sector in London in terms of employment, GVA, productivity, business investment and stock since 2012.

## *Economic benefits*

Findings indicate the objectives of London Tech Week are being met, in that the event is being seen as building London's reputation in the tech industry across the globe and successfully positioning London as the tech capital of Europe. It is also contributing to driving investment into the city and employment in the tech sector by helping create leads for setting up and expanding businesses.

The main beneficiaries of this business event in terms of GVA (short-term economic benefit) are venues, hotels and the hospitality sector. Other industries and businesses will in turn benefit as a result of the multiplier effect. Over the longer term the beneficiaries will be the tech companies in London and across the world as well as the tech workforce/talent. As the tech industry develops and grows in London, the potential for 'whole of city' economic benefit will also increase.

**Table 4.4:** Development of the tech sector in London by selected indicators (2012-2017)

| Digital technology, London | 2012 | 2013 | 2014 | 2015 | 2016 | 2017 (forecast) |
|---|---|---|---|---|---|---|
| Employment (000's) | 166.6 | 178.1 | 206.4 | 221.3 | 240.0 | 242.7 |
| GVA (£m, 2013 real prices) | 16,220 | 16,773 | 18,428 | 20,413 | 22,503 | 23,398 |
| Productivity (£, 2013 real prices) | 97.4 | 94.2 | 89.3 | 92.2 | 93.7 | 96.4 |
| Business stock (000's) | 32.0 | 33.7 | 37.6 | 41.9 | 46.4 | 47.4 |

4

## *Tech industry and destination reputation benefits*

London Tech Week was seen to send a positive message worldwide about London's booming tech sector and digital industry. Data obtained via social media research tool, Crimson Hexagon, revealed how people who engaged with London Tech Week 2017 via social media felt about the London tech scene.

By analysing social media posts that mentioned London's tech scene/sector/industry between 1 November 2016 and 30 June 2017, an understanding of how the business event impacted sentiment amongst attendees was garnered. More than 40,000 social posts mentioning London's tech sector were tracked, and of these 67% were positive (as opposed to neutral: 9% or negative: 24%). These positive posts praised organisers, or promoted the event or the tech sector as a result of the commentator's engagement with the event (attending events, reading articles, hearing comments from others etc.). As with all social campaigns, there are always social posts considered negative. The social posts identified as negative in this research were not so much about the tech scene in London or the business event itself, but more about the wider economic and political drivers behind the sector and the industry.

Positive posts included comments relating to:

- The fact that the tech sector is thriving in London;
- The way the sector is contributing to London's economy;
- The business event being a celebration of the sector;
- The future of the sector being bright;
- The fact that #LondonIsOpen for business.

Such comments were seen by the organisers as reaffirming the positioning of London as the best place for investment, talent and innovation in the tech industry. During the specific week of the business event, the proportion of positive posts about London's tech scene increased to 75% (14,854). The following are some of the most noted comments.

4

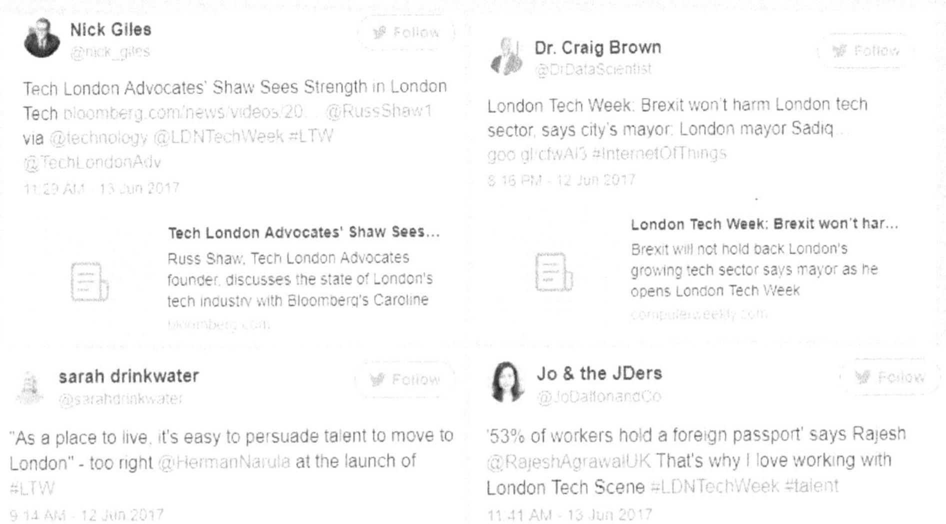


These positive social messages related to London Tech Week, improving London's image as a destination and helping to showcase the booming tech industry to a wider audience. The overall social reach of London Tech Week 2017 was over 121 million people (between 1 November 2016 and 30 June 2017), and with much of this reach reflecting positive commentary, it can be presumed that the organisers' objectives around conveying a positive message about London were largely met.

## Implications for future research

Over the four years the business event was run (as at 2017), the key lessons learned were around:

- ☐ The importance of working closely with relevant stakeholders across the city (including government bodies, both local and national)
- ☐ Having strong, connected ambassadors
- ☐ Developing good working partnerships
- ☐ Clearly defining the curation and content
- ☐ Being clear in advance as to the outcomes they wished to achieve (to ensure the event did not spread and weaken the objectives)
- ☐ Being aware that the legacy and outcomes were key (things like reputation building, attracting crowd-sourced events, animating the city and generating longer term foreign direct investment, as well as attracting and retaining talent, rather than simply focussing on short-term economic benefits).

Whilst purely economic benefit data (as provided in this case study) is of use, more long-term, qualitative data would add considerable depth and credibility to the overall evaluation of how an event such as this was able to impact economic growth in a city like London. It would also enable event organisers, industry peak bodies and governments to leverage economic benefit legacies and grow identified industry sectors where needed. When data collection is planned into the various stages of an event, organisers are better placed to evaluate the successfulness of their strategic approaches.

## Conclusions

In summary, this case study has reported on results from both the 2016 and 2017 London Tech Week events. The findings of the research reveal:

**For London Tech Week 2016:**

- The GVA calculated was £2,531,752
- The overall economic benefit calculated was £21,218,425
- The event generated at least 10 leads for London & Partners' Foreign Direct Investment team, with estimates of a total of at least 163 jobs being created in the first year and 294 within three years of the business event.

**For London Tech Week 2017:**

- The GVA calculated was £3,615,721;
- The overall economic benefit calculated was £45,915,245;
- The business event generated at least 23 leads for London & Partners' Foreign Direct Investment team, with a total of at least 24 estimated jobs created in year 1 and 51 in year 3;
- Of the social media messages generated by the business event in relation to London's tech scene, 67% of conversations were positive. This figure rose to 75% during the week of the business event itself.

The findings also indicate that, as a result of London Tech Week, the following benefits were realised for the people and the economy of London, and (more specifically) for its technological and digital industries:

1. The London Tech Week business event added value to London as a destination for tech talent, innovation and investment.
2. The event showcased the best of the vibrant, growing and innovative tech sector in London and created opportunities for technology businesses and workforce growth.
3. Positive growth and development of the tech industry in London occurred as a result of the business event.
4. The event triggered positive conversations about the tech sector in London and helped push that positive message to an international level.
5. Value was added to London's economy (direct and indirect) from short-term tourism and the multiplier economic impacts of hosting this business event.
6. The business event helped build London's reputation in tech across the globe and put London on the map as the tech capital of Europe.

These indications are based on secondary financial data supplied by London & Partners and their agencies, and other statistical reports. There is no doubt that Tech Week continues to fulfil economic development and destination reputation objectives for London and be seen globally as an impressive celebration of the host country's tech industry.

Importantly, London Tech Week connects investors and larger enterprises at the forefront of digital transformation with tech start-ups and scale-ups looking to expand or relocate to London. As a result of the event, the start-ups have a platform (exhibition space) through which to showcase and pitch their ideas and capacity, and seed investors are given the opportunity to evaluate the potential they see in the start-ups for future growth. In this way, the business event acts as a catalyst for ongoing tech industry growth and contributes to the ongoing economic development of London.

Further research, incorporating primary and qualitative data related to attendee experiences, would enable a more complete picture of the event's success to be established.



## Theoretical discussion

Traditionally, the economic contributions of business events have been tied to expenditure associated with attendees, in particular those visiting from outside a local region. Thus, the tourism contribution has been the standard benchmark (internationally and domestically) for measuring the economic scale of a business event (Jago, 2012).

This focus on tourism expenditure to assess the value of business events is not surprising when one considers that the entities charged with planning and delivering business events are often a part of government tourism agencies (Weber *et al.*, 2011) and, hence, have a tourism focus. However, it is important to understand that the tourism contribution is merely the tip of the iceberg in terms of the overall economic benefits business events can bring to host economies.

More recent research (Foley *et al.*, 2014; Edwards *et al.*, 2017) has identified a broad range of business event outcomes and legacies beyond tourism which have direct impacts upon industry associations, government authorities and the wider community. Many of these impacts are likely to result in significant short and long-term economic benefits, suggesting that the real economic benefits of business events are much greater than previously thought.

## Economic benefits generated by business events

Foley *et al.* (2010; 2013) and Edwards *et al.* (2016) identified a number of ways in which business event impacts can result in economic benefits far beyond those generated by delegate expenditure.

For governments, business events:

- Play a key role in assisting the implementation of government initiatives (e.g. the Australian Government's National Innovation and Science

Agenda is designed to *drive the generation of smart ideas that create business growth, local jobs and global success*). Thus conferences provide the opportunities necessary for personal contact and formation of networks which drive collaboration and innovation.

- Provide opportunities to bring about social change (this may relate to health, economic, environmental or other social issues). The business event forum facilitates generation of ideas and collaborations which can bring about solutions more cost effectively. It can also provide an effective platform for broader discussion/debate/communication around the required change, all of which bring cost savings in terms of marketing, communication, public relations, administration and collaboration).
- Help shape and advance a country's economic development strategies (providing forums for interaction between global investors and local businesses).
- Enable the local sector to showcase its abilities and achievements to a global audience (thereby generating opportunities for foreign trade and investment in research and business).
- Increase a nation's expertise (not only as a result of having exposure to the world's experts and new knowledge, techniques and technologies, but also because the local sector is stimulated and inspired to rise to the challenge of the global competition).
- Support a destination's tourism and hospitality industry growth, as well as secondary businesses as a result of multiplier impacts.

For professional associations, business events:

- Contribute to better international collaboration by facilitating the formation of global research and business networks (which can lead to cost-saving partnerships, trade agreements, resource sharing and significant economic benefits for those involved).
- Provide delegates with new ideas, skills and knowledge (which can be used to advance their field or practice, improve competitive advantage for their employers, or develop innovative cost-saving practices).
- Promote local sector expertise to an international audience (and hence open up opportunities for economically beneficial collaborations and partnerships to be formed).
- Attract international expertise to a host destination (creating inbound migration or foreign investment in the case of company relocation).
- Raise the profile of an organisation, issue or association to generate support from within government, the community or the media (which can result in increased funding).
- Provide valuable opportunities for businesses to find new customers and contacts for trade (resulting in increased sales and potential profits).

- Facilitate connections between research and business entities (assisting the translation of research and innovation into commercial outcomes and hence profit generation).
- Enable associations and groups to generate income for use in other cause-related projects.

For the community, business events:

- Encourage collaborations which expose people to new contacts, ideas and knowledge (which often brings innovation, new business, upskilling or training).
- Build on the strength of the host destination by adding skills to the local industry.
- Drive social change, which can have important economic benefits (for example, conferences which focus on health-related issues like smoking or obesity can drive social behavioural changes which can result in significant savings in terms of reduced national health costs).
- Bring host destination benefits as a result of increased local employment and patronage (which can energise a community and introduce new money into the area).
- Boost ongoing tourism in a region as a result of repeat visitation (i.e. where attendees may return post-conference for purely tourism purposes, often bringing others with them).
- Raise the profile of a destination, giving the destination a reputation in the global marketplace and adding to the destination's brand (which can also attract new businesses and foreign investment).



One of the many examples of how business event outcomes have translated into significant economic benefits is that of the Australian Royal Mint (the approved primary producer of Australia's currency), as cited by the Business Events Council of Australia (2015). In 2010 the Mint hosted the biannual Mint Directors Conference (MDC) in Canberra, Australian Capital Territory (ACT), with 350 industry representatives from around the world attending. The tourism-related economic benefit delivered from the conference was estimated at A$425,000 (for the ACT economy) and A$650,000 (for the wider Australian economy). These amounts were calculated using traditional tourism contribution formulas. However, these economic contributions were relatively minor compared to the longer term economic benefits which resulted from the event.

Prior to the 2010 MDC, the Mint had minimal export sales (approx. $A2m per annum), but after the event, sales grew exponentially. As part of the conference, all attendees were given a tour of the facility and information on how it conducted its business. The significant ensuing growth in export sales has been directly attributed to the hosting of the 2010 MDC.

> *As a result of the conference, and the exposure of the Royal Australian Mint … (we have) entered into several export contracts resulting in a total revenue value*

*of over $12 million per annum, with some of these being multi-year contracts of over 10 years... Hosting the MDC enabled us to bring the decision-makers to where the work is done, in Canberra at the Mint. No other marketing or sales initiative would have had the same immediate and significant impact.*
(Dr Prabir De cited in Business Events Council of Australia, 2015: 12).

Having the MDC in Canberra also delivered other benefits to the Mint and its staff, including in skills development and in expanding the views of staff on what can be achieved.

As is demonstrated by the experience of the MDC, attracting appropriate business events represents an excellent pathway to demonstrating local expertise and growing business.

## The Business Events Value Triangle

Given the range of benefits identified above, it is becoming increasingly evident that tourism expenditure is merely a sub-set of the potential total economic value of business events. Advanced by Deery *et al.*, 2005, the business events triangle illustrates the different components of value in the business events sector, from its initial tourism contribution right through to benefits with the potential to create significant, long-term economic advantage. This model provides a more realistic conceptualisation of the proportion of economic value generated by tourism and delegate expenditure, when compared to the value which can be created in other ways. It also demonstrates why a focus on tourism contributions alone will result in not fully or properly accounting for the vast majority of other contributions business events may generate.

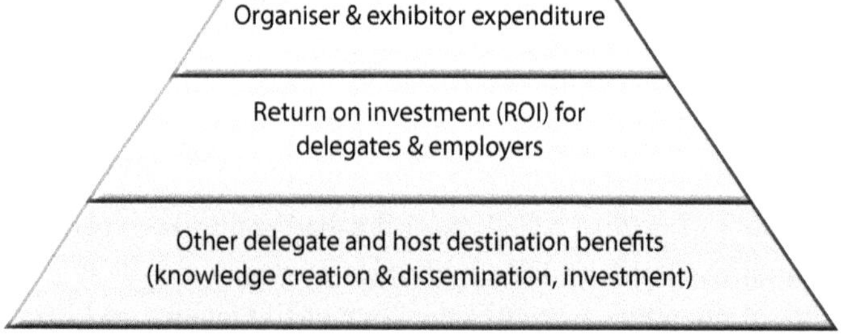


**Figure 4.1:** The Five Components of Business Events Value

## Ways of measuring the economic benefits of events

As consensus that business events have the capacity to generate significant economic benefits beyond attendee spend grows, so too do questions around:

- How such benefits can be properly planned for (leveraged); and
- How such benefits can be tracked, measured and included in post-event evaluations.

Many non-tourism benefits are indirect (i.e. difficult to be directly and exclusively linked to something that happened at the event) and occur over a longer term (perhaps 5–20 years post-event). Similar issues have long troubled practitioners in arts, leisure and other service provision. How does one measure the economic impact of providing museums, art galleries or public playgrounds for example (see Throsby, 1994; Scott, 2006; Böhm & Land, 2009; Leggett, 2009; Bakhshi *et al.*, 2015)? How can the full economic value of having clean beaches, waterways and national parks be measured (Marikan et al., 2006; Tisdell, 2006)? And how can provision of free community health care be shown to be an economically valid expenditure of public money (Walker, 2005; Whitely, 2006)?



In fact, there are many important facilities and services enjoyed by developed communities today which may not generate direct, immediate economic benefit, yet generate impacts considered to be economically rational. If free education was not provided, a country's future generations would be unskilled and reliance on imports and foreign labour would see the country's GDP fall. If beaches or national parks were left uncared for (or sold to private entities), local tourism income would fall and amenity for residents would drop, likely leading to a number of costly social problems. A number of formulas and processes to rationalise and measure the real economic value of providing or supporting these types of facilities and services have been developed (and continue to be developed). With the enormous capacity for business events to support business, connect entities, generate information and knowledge, and support innovation and new ideas, it is now time for similar economic impact and measurement tools to be developed and tailored specifically for the industry.

An economic impact framework explicitly addressing the convention sector was developed by Dwyer and colleagues in the late 1990s (Dwyer & Forsyth, 1997; Dwyer *et al.*, 2000). The framework focused on ways to estimate business events' economic contribution to a host destination based on previous methods of estimating the economic contribution of special events (e.g. Burns & Mules, 1986). It also formed the basis of the guidelines established by the United Nations World Tourism Organization (UNWTO, 2006), which were subsequently adopted for numerous studies aiming to establish the economic contributions of events to national economies in Australia, Canada, United States, Mexico and Denmark (Deery *et al.*, 2005; Mauritz, 2008; PwC, 2011a, 2011b; Visit Denmark, 2012).

In their discussion of the framework, Dwyer et al. (2000) made mention of things like the economic contributions made by conferences in terms of:

- Exchanging ideas;
- Continuing education;
- Maintaining professional contacts;
- International exposure for the host destination;
- The merits of drawing specialists in a field together;
- Triggering private and public development;
- Growth of existing businesses;
- Establishment of new businesses;
- Links between nations and firms;
- Opportunities for international cooperation;
- Promotion of national interests;
- Increased business confidence and activity; and
- Improved property values (as a result of construction and development).

However, the framework itself focused predominantly on contributions made by events to the visitor economy and direct expenditure. Development of a framework capable of estimating and measuring these more intangible economic contributions remains elusive. Fortunately, progress is being made. Following years of research into delegate perceptions of conference outcomes, Edwards *et al.* (2017) related the experiences of a number of notable Australians who felt business events had made significant contributions to their professional success. The stories included examples of how business events had led to:

- Scientific discoveries that changed the course of research (indirectly averting the wastage of millions of dollars in funding);
- Research connections forged at conferences which had directly resulted in development of vaccines and other medical discoveries (saving hundreds of thousands of lives globally each year and billions of dollars in associated health care costs);
- Development of innovative farming techniques (creating whole industries which previously did not exist, but which are now thriving).

Whilst aspects of the data supporting this research was predominantly qualitative in nature, the evidence is compelling. It is also useful in that it provides a template for how other such outcomes might be captured and incorporated into tools and a comprehensive economic framework in the future.

Perhaps the business events industry can draw upon examples from other areas of the events industry that have been reconfigured to plan for, leverage and measure long-term and often economically significant benefits for host destinations. The Olympic Games, for example, was once held for the sporting event itself, and success was measured in terms of things like tickets sold, tourism expenditure, and sponsorships, advertising and broadcast rights. When concerns emerged that some hosts did not achieve the economic returns anticipated and that built infrastructure was under-utilised post-event, there was a shift in focus. Bid proposals which showed evidence of legacy planning likely to bring long-term economic and social advantages to host destinations were given preference, thereby enhancing their value proposition.

London 2012, for example, planned from the outset to orient the Games toward a legacy of regeneration, growth management and sustainability to tackle the social problems in the city's boroughs, considered to be one of the poorest and most deprived areas in Europe (OECD, 2010). The organisers strategically planned delivery of the event to create:

- A lasting environmental legacy with world-class standards in sustainable construction and amenity;
- Public transport infrastructure upgrades to support long-term city connectivity;
- Regeneration frameworks;
- A program to improve the competitiveness of SMEs and supply chains in response to Olympic-related opportunities.



In a report on the developmental benefits possible from staging such events, the OECD refers to the importance of vision and leadership in the creation of legacies, stating that legacy planning has to be vision-led and "should span a series of events and opportunities, of benchmark moments and catalysts, rather than a single major event" (OECD, 2010). Such legacy planning should:

- Include both short-term and long-term initiatives;
- Focus on direct and indirect impacts;
- Be based on defined and specific programs;
- Address wider institutional and organiser benefits.

The report highlights that legacy planning should have robust leadership and dedicated resources and skills *separate to* the resources being utilised to host the event, because the skills needed to create ongoing and lasting legacy benefits are different from those required to stage an event (OECD, 2010). Long-term legacy impacts rarely happen automatically, and the skills required to derive and optimise such benefits are usually developed over time. However, the report also stressed that developing ways to measure the relationship between events and the legacy impacts they created was crucial, stating:

> *"The problem of attribution is significant. How to count impacts that are direct, indirect, assisted, or facilitated by the event effectively remains a challenge."*

The London Tech Week case study is useful, in that whilst it incorporates methods and processes commonly used to measure direct, short-term and immediate expenditure, organisers clearly have their sights set on the end game and have implemented strategies which attempt to capture at least some of the long-term economic outcomes as well. The event, at its core, is designed to create long-term growth and economic development for London by:

- ☐ Stimulating, developing and expanding the London tech industry;
- ☐ Bringing exciting new talent to the industry from overseas;
- ☐ Bringing new business ventures to London (foreign investment);
- ☐ Enhancing and cementing London's reputation for excellence in the tech industry to support growth and development into the future.

These are lofty goals and there is no doubt it would be difficult to accurately capture or quantify the full extent to which such outcomes were being achieved by the event. For example, what if a talented New Zealand tech wizz, visiting her brother in London, hears about London Tech Week in the media and drops in on a couple of pop-up events. Five years on she moves to London, mainly because she is missing her brother, but also because of what she saw at the event. She gets a great job in the tech industry and eventually sets up a successful international business. But was it the event or her brother which prompted the move, and how could organisers of the event capture her in their data if she didn't specifically register at the event or let them know once she had relocated? To complicate things further, many of these goals may take years to eventualise. An entrepreneur at the 2017 event may be so struck by the buzz that he makes a decision then and there to launch his start-up in London, but he doesn't have the capital or the family situation to make it a reality until 2029. How is this attraction of talent captured? It may be even more problematic when attempting to track things like the reputation of an excellence hub; something that may have taken a decade or more to be formed. Hence, even if appropriate methods *were* formulated, data would need to be tracked and captured for up to 20 years before one could have confidence it reflected the full picture. And that is just the first set of issues. From there, it is even more challenging to accurately measure the *economic benefit* those outcomes generated.

But does that mean such outcomes and benefits are any less significant or valid? And even more importantly, do the inherent challenges of carrying out such research mean we should not try? Whilst measuring visitor spend and tourism generation is important, it is development of tools and practices that will enable the business events industry to accurately track, measure and leverage the full extent of the economic benefits it creates that will safeguard its future.

## Contributions to the visitor economy

Whilst the issues raised above revolve around long-term economic outcomes from business events, there is no doubt that the short-term outcomes and benefits associated with visitor spend also need to be accurately measured. Short-term economic impacts may be smaller, but they are more immediate and (for now at least) more objectively measurable, making them useful for leveraging to decision-makers, engaging stakeholders and building support. The case study provides some helpful approaches around measuring certain event outcomes in the short term, beyond visitor spend, but it also includes this important aspect of event evaluation.

Central to measuring the contribution of events (including business events) to the visitor economy is the identification of inscope expenditure (Jago & Dwyer, 2006). Such expenditure represents an injection of 'new money' into an area by delegates and accompanying persons who would not have spent in the destination had the business event not been held. It is important to establish that the business event is the main reason for the trip for visitor expenditure to be regarded as 'new money'.

Several factors influence inscope expenditure. These include the:

- Number of delegates and accompanying persons, and their daily expenditure;
- Types of delegates and types of conventions;
- Trip duration/length of stay;
- Costs at the business event destination.



Estimations of inscope expenditure attributable to a business event do not include expenditure by delegates who originate from within the host destination (apart from those who are counted in retained expenditure) (Jago & Dwyer, 2006; Foley *et al.*, 2014b; Foley et al., 2016). It also does not include any return on investment (ROI) for sponsors and exhibitors (Edwards *et al.*, 2011). And whilst an increasing number of business events have begun trying to quantify such economic impacts, there are challenges around sponsors and exhibitors being unwilling or unable to either directly track their ROI from particular business events or to disclose such sensitive information publicly.

In the past. some studies adopted incorrect assumptions which inflated the importance of business event contributions. For example, some of the early impact assessments incorrectly counted gross expenditure rather than only 'new money' (e.g. Braun, 1992), including all purchases of final goods and services linked with the business event by the various interested parties (delegates, accompanying persons, other convention participants, organisers and sponsors), regardless of their origin. Similarly, incorrect treatment of expenditure on airfares potentially inflates business event contributions when portions of that expenditure not received in a particular region are nevertheless included in impact assessment studies, or deflate it if incorrectly excluded. Johnson *et al.* (1999), in a study for the Bureau of Tourism Research (BTR) on the economic contribution of conventions to Australia, correctly included expenditure on international airfares where this expenditure was received by an Australian business. In contrast, the earliest study on the industry's economic impact by the Australian Tourist Commission (Australian Tourism Commission, 1990) excluded international airfares from expenditure items. Dwyer and Forsyth (1997) argued that this exclusion appeared unjustified since a substantial number of delegates used the country's international carrier. In recent decades, the complex international ownership of major airlines has made it difficult to attribute airfare expenditure to a particular region and airfares are now generally excluded altogether from inscope expenditure reporting.

Expenditure within the destination by organisers, associations and sponsors also represents an injection of 'new money' if these parties are located outside the expenditure zone, and thus should be added to estimates of total inscope expenditure associated with the business event. Revenues received by business event organisers (including registration fees, corporate sponsorship, government sponsorship and sponsorship from associations) should not be included in inscope calculations. These types of revenues are typically re-spent by organisers within the destination on items such as venue hire, equipment, various business expenses, catering and social programs, and to avoid double counting, revenue (including registration fees) should not be included in calculations of business event visitor expenditure.

There are several additional considerations in accurately establishing inscope expenditure, namely the treatment of expenditure by local event attendees, switched expenditure and direct imports (Dwyer et al., 2000; Jago & Dwyer, 2006; Jago, 2012).

## Transferred expenditure

Expenditure by local event delegates should be excluded from economic impact assessments as its inclusion distorts and grossly overstates the economic impact of a business event. Rather, such expenditure represents a transfer of expenditure from one expenditure category to another as locals would have purchased goods and services in the locality, with local businesses still benefiting from demand regardless of the event. In a similar vein, sponsorship from local sources should also be regarded as transferred expenditure unless it represents an additional injection of funds.

## Retained expenditure

Conversely, expenditure of local attendees (and that of exhibitors and sponsors located within the destination) who would have attended the event anyway, even if it had been held in another location, can be considered inscope expenditure. This element of inscope expenditure is often ignored, but it may be particularly relevant to many conferences. For example, many members of professional associations are inclined to reliably attend the annual event of their association no matter where it is held. Such retained expenditure prevents the outflow of money from an economy and justifies its inclusion in the calculation of inscope expenditure.

## Switched expenditure

The expenditure of attendees who adjusted the timing of their visit to a destination to attend a business event, but would have visited the destination regardless of whether or not the business event occurred, should not be attributed to the business event. Thus, it is important to establish whether the timing of the attendee's visit was changed to coincide with the business event. The same principle applies to the expenditure of exhibitors and sponsors. If their expenditure would have occurred in the destination in any case, but there has simply been a switch in the timing of this expenditure to coincide with the business event, then this expenditure should not be included as inscope expenditure.

### Direct imports

Imports of goods and services directly related to a business event from outside the region represent a leakage out of the host economy. Smaller regions within an economy tend to be less self-sufficient in producing goods and services and therefore have a higher overall propensity to import. Expenditure on goods and services outside of the region should not be included in inscope expenditure.

### Allocation of aggregate inscope expenditure and multiplier effects

Both volume and pattern of expenditure of business event visitors and organisers is of key relevance to the economic impact assessment of business events since different industries have different types of linkages with other sectors in the economy, and hence different multiplier effects. Inscope expenditure can be allocated to individual industry sectors on the basis of expenditure surveys of business event visitors and organisers. Several key industry categories capture the main types of business event-related expenditure, namely accommodation, transportation, shopping, food and beverage, entertainment, and organised tours (Dwyer & Forsyth, 1997).



As inscope expenditure has flow-on effects (indirect plus induced) to an economy, various multipliers are used to determine the contribution of business events to destination output, value added, and employment (Jago & Dwyer, 2006). A recent report from the Business Events Council of Australia (2020) provides a comprehensive overview of the indirect contributions of business events.

### Practicalities of evaluating contributions of business events to the visitor economy

In a review and critical analysis of various economic impact assessment studies undertaken around the world, Jago (2012) identified similarities and differences among studies undertaken at the national and sub-national levels. At the national level, he assessed five studies undertaken at that time in the US, Mexico, Denmark, Australia and Canada, and noted that there was consistency among the studies as they all adopted the UNWTO (2006) guidelines. Since the UNWTO guidelines are based on sound economic principles, these studies produced internationally credible results. Yet, all these require substantial resources and thus, potentially, they would be difficult to undertake on a regular basis for comparative purposes.

## *Economic contributions beyond the visitor economy*

Despite the issues of resourcing mentioned above, methodologies for estimating contributions to the visitor economy are well established and widely used (Dwyer *et al.*, 2000; Jago & Dwyer, 2006; Jago, 2012; Foley et al., 2014b; Foley et al., 2016; Foley et al., 2019). Research and methodologies for estimating the economic contributions of business events *beyond* the visitor contribution, however, are less well established.

### Investment and new business

As discussed earlier in this chapter, in addition to the contributions linked to inscope expenditure (new money, jobs, multiplier effects), business events provide opportunities for destination enterprises to attract foreign direct investment and new business. These important economic benefits contribute to national economies, research and industry sectors, and local enterprises. The case study refers to multiple instances of foreign direct investment into the London tech industry as a result of London Tech Week. An earlier study (Foley *et al.*, 2010; Foley *et al.*, 2013) notes significant research investment into AIDS research following an international AIDS conference in Sydney. Further research (Edwards *et al.*, 2016) has found that business events provide exhibitors at conferences and trade shows with new business networks and leads.

### Knowledge economies and upskilled workforces

The economic benefits of business events also include contributions to knowledge economies and the upskilling of workforces. Many of the chapters in this book make reference to both of these phenomena. When a business event comes to a destination, it brings cutting-edge knowledge, techniques and technologies to the relevant industry sectors (for examples, see Chapters 3 and 5). This allows industries to function more efficiently and effectively, while the people who work in the related industries upskill and their communities reap the benefits.

## Lessons learned

This chapter presented a case study and a range of theoretical perspectives that underscore the importance of accurately establishing the economic contributions of business events following established guidelines to ensure not only accuracy but also comparability. Accurate and consistent approaches to impact measurement are especially vital since decisions about resource allocation by both private and public sector stakeholders greatly depend on accurate information regarding potential gains. Information on the sector's impacts is essential to determine the appropriate extent of government support through initiatives such as development loans, investment incentives, tax concessions, training assistance, favourable visa access for international delegates, and promotional and marketing campaigns (Cooper, 1999; Rogers & Davidson, 2016). In the absence of accurate data on the industry's full economic significance, destinations may under or over-allocate resources to the business events industry, resulting in reduced net economic benefits.

Having a critical mass of key decision-makers in a destination for a large-scale business event clearly facilitates active solicitation of those decision makers' support. Destination representatives can take advantage of the opportunities accorded by large business events to make new contacts within a particular sector or meet with existing contacts to offer assistance in the set-up or expansion of their business in the destination, including future job creation. Furthermore, having like-minded people from a particular industry sector is likely to create substantially more 'buzz' on important issues, leading to ideas and solutions. As indicated in Chapter 8, attaining a 'hub' status in an industry sector facilitates the strengthening of local-global knowledge networks, further catalysing innovation and momentum in that sector. Few contest that such business event outcomes lead to economic benefits, but it is the issue of capturing, tracking and measuring the extent of those benefits that is the challenge. Without robust evidence, and with an increasing range of events laying claim to being catalysts for such economic growth, it is those industry players who proactively plan for, track and measure their broader economic achievements who will have evidence to leverage support for their events.

Business events catalyse investment into research, fire business growth, and promote knowledge-sharing and adoption of cutting-edge ideas, technologies and techniques into professional practice. For many decades, estimations of the economic contributions of business events to destinations have been marooned on the narrow reef of the visitor economy. These contributions are substantial and should not be ignored. However, they are just the tip of the iceberg. More research is required, but it is already evident that business event contributions beyond tourism expenditure have the capacity to generate significant legacies for host destination innovation, business growth and economic development.



## References

Australian Tourism Commission (1990) *Convention Delegate Expenditure Study*, prepared by Functional Research and Marketing, February 1990.

Bakhshi, H., Fujiwara, D., Lawton, R., Mourato, S. and Dolan, P. (2015) *Measuring economic value in cultural institutions*, Arts and Humanities Research Council.

Böhm, S. and Land, C. (2009) No measure for culture? Value in the new economy, *Capital & Class*, **33** (1), pp. 75-98. https://doi.org/10.1177/030981680909700105

Braun, B. (1992) The economic contribution of conventions: The case of Orlando, Florida, *Journal of Travel Research*, **30** (3), 32-37.

Burns J. and Mules T. (1986) An economic evaluation of the Adelaide Grand Prix, in G. Syme, B. Shaw, P. Fenton and W. Mueller (eds), *The Planning and Evaluation of Hallmark Events*, Aldershot, England: Avebury, pp. 172-185.

Business Events Council of Australia (2015) The value of business events to Australia. https://www.businesseventscouncil.org.au/files/View_Report.pdf

Business Events Council of Australia (2020) The value of business events to Australia. https://www.businesseventscouncil.org.au/files/Value_of_Business_Events_2018-19.pdf

Cooper, M. (1999) Prediction and reality: The development of the Australian convention industry 1976-1993, and beyond, *Journal of Convention & Exhibition Management*, **1** (4), 3-15.

Deery, M., Jago, L., Fredline, E. and Dwyer, L. (2005) N*ational Business Events Study*, Canberra: BTR.

Dwyer, L. and Forsyth, P. (1997) Impacts and benefits of MICE tourism: A framework for analysis, *Tourism Economics*, **3** (1), 21-38.

Dwyer, L. Forsyth, P. and Spurr, R. (2005) Estimating the impacts of special events on the economy, *Journal of Travel Research*, **43** (May), 351-359.

Dwyer, L., Mellor, R., Mistilis, N. and Mules, T. (2000) A framework for assessing 'tangible' and 'intangible' impacts of events and conventions, *Event Management*, **6** (3), 175-189.

Edwards, D., Foley, C. and Hergessel, A. (2016) *Conferences: Catalysts for thriving economies*, report prepared for Business Events Sydney, UTS, Sydney. https://www.besydney.com.au/media/1611/catalysts-for-thriving-economies-research.pdf

Edwards, D., Foley, C. and Schlenker, K. (2011) *Beyond tourism benefits: Measuring the social legacies of business events*, report prepared for Business Events Sydney, UTS, Sydney.

Edwards, D., Foley, C. and Malone, C. (2017) *The Power of Conferences*, UTS E-Press. https://utsepress.lib.uts.edu.au/site/books/m/10.5130/978-0-6481242-0-7/

Foley, C., Edwards, D and Hergessel A. (2016) *Asian incentive events in New South Wales: Expenditure and retail impact*, report prepared for Business Events Sydney, University of Technology Sydney. Available at: https://www.besydney.com.au/media/1616/asian_incentive_events_nsw.pdf

Foley, C., Edwards, D and Hergessel A. (2019) *Direct expenditure attributed to international conferences held in New South Wales 2017-2018*, report prepared for Business Events Sydney, University of Technology Sydney. Available at: https://www.besydney.com.au/media/3360/bes002_de6_conferencesinfographic_a4-2.pdf

Foley, C., Edwards, D. and Schlenker, K. (2014a) Business events and friendship: leveraging the sociable legacies, *Event Management*, **18** (1), 53-64.

Foley, C., Edwards, D., Schlenker, K. and Hergessel, A. (2014b) *Estimating inscope expenditure for business events held in New South Wales*, report prepared for Business Events Sydney, UTS, Sydney. Available at: https://www.besydney.com.au/media/1617/estimating-inscope-expenditure-attributed-to-business-events-held-in-nsw.pdf

Foley, C., Schlenker, K., Edwards, D. and Hayllar, B. (2010) *A scoping study of business events: beyond tourism benefits*, report prepared for Business Events Sydney, UTS, Sydney. Available at: http://hdl.handle.net/10453/16848

Foley, C., Schlenker, K., Edwards, D. and Lewis-Smith, L. (2013) Determining business event legacies beyond the tourism spend: An Australian case study approach, *Event Management: An International Journal*, **17** (3), 311-322.

Foley, C., Edwards, D and Hergessel A. (2016) *Asian incentive events in New South Wales: Expenditure and retail impact*, report prepared for Business Events Sydney, University of Technology Sydney. Available at: https://www.besydney.com.au/media/1616/asian_incentive_events_nsw.pdf

Jago, L. (2012) *The value of business events*, report prepared for Tourism & Business Events International, Melbourne.

Jago, L. and Dwyer, L. (2006) *Economic Evaluation of Special Events: A Practitioner's Guide*, Altona, Australia: Common Ground Publishing.

Johnson, L. Foo, L.M. and O'Halloran, M. (1999) *Meetings make their mark: characteristics and economic contribution of Australia's meetings and exhibitions sector*, report prepared for the Bureau of Tourism Research, Canberra.

Legget, J. (2009) Measuring what we treasure or treasuring what we measure? Investigating where community stakeholders locate the value in their museums, *Museum Management and Curatorship*, **24** (3), 213-232. DOI: 10.1080/09647770903073052.

London & Partners (n.d.) How we measure our impact, retrieved 14 July 2017 from http://londonandpartners.com/about-us/how-we-measure-our-impact

London Tech Week (2017) Mission statement, retrieved 14 July 2017 from https://londontechweek.com/

Marikan, D., Radam, A. and Siti Baizura, J. (2006) *The economics of recreational park conservation: A case study of Bako National Park*, Staff Paper 4/2006, Faculty of Economics and Management, University Putra Malaysia, Malaysia.

Mauritz (2008) *The economic contribution of meetings activity in Canada*, report prepared for MPI Canada.

OECD (2010) *Local development benefits from staging global events: Achieving the Local Development Legacy from 2012*.

PwC (2011a) *The economic significance of meetings to the US economy*, report produced for the Convention Industry Council.

PwC (2011b) *The economic significance of meetings to Mexico*, report produced for Tourism Mexico.

Rogers, T. and Davidson, R. (2016) *Marketing Destinations and Venues for Conferences: Conventions and Business Events*. Routledge.

Scott, C. (2006) Museums: impact and value, *Cultural Trends*, **15** (1), 45-75. DOI: 10.1080/09548960600615947

Thiebault, A. (2018) *London Tech Week*, London & Partners. https://www.themeetingsindustry.org/web/wp-content/uploads/JMIC_Case-Study_London-and-Partners_London-Tech-Week-17.12.pdf

Throsby, D. (1994) The production and consumption of the Arts: A view of Cultural Economics, *Journal of Economic Literature*, **32** (1), 1-29. Retrieved 4 April 2021 from http://www.jstor.org/stable/2728421

Tisdell, C. (2006) Valuation of tourism's natural resources, in: L.Dwyer & P.Forsyth (eds.) *International Handbook on the Economics of Tourism*, p. 359, Edward Elgar Publishing.

UNWTO (2006) *Measuring the Economic Importance of the Meetings Industry: Developing a Tourism Satellite Account Extension*, Madrid, Spain.

Visit Denmark (2012) *The economic contribution of meeting activity in Denmark*, report prepared for the International Congress and Convention Association.

Walker, D. and Jan, S. (2005) How do we determine whether community health workers are cost-effective? Some core methodological issues, *Journal of Community Health*, **30**, 221-229. https://doi.org/10.1007/s10900-004-1960-4

Weber, K., Chon, K and Chan, A. (2011) *Excellent CVBs of the Future*, Hong Kong Polytechnic University Working Paper, Hong Kong.

Whitley, E., Everhart, R. and Wright, R. (2006) Measuring return on investment of outreach by Community Health Workers, *Journal of Health Care for the Poor and Underserved*, **17** (1), 6-15. doi:10.1353/hpu.2006.0015

# 5 Organisational Development Depends on Knowledge Sharing

*Because of the range of legacies interviewees reported had occurred, we recommended that other local host committees initiate and document similar legacy outcomes where possible, in order to inspire delegates to make the most possible use of their participation. Those in the public and private sector who are involved in organising or hosting business events should be more than willing to support such initiatives because it will add a broader societal purpose to their function apart from the marketing of meeting facilities.*

André Petzold, Former Business Development Manager, Destination Jönköping, Sweden

The direct economic impacts of conferences have been a focus of scholarly interest for decades due to the benefits associated with tourism, trade and investment to the host city (Braun, 1992; Rutherford & Kreck, 1994; Grado *et al.*, 1997; Foley *et al.*, 2016). However, less tangible benefits, such as the return on personal investment for attendees (Dwyer *et al.*, 2000), have received far less attention (Mair & Jago, 2010; Foley *et al.*, 2013; Mair, 2013; Mair & Frew, 2018; Edelheim *et al.*, 2018).

In fact, it is becoming increasingly apparent that benefits (both short and long term) beyond direct economic outcomes make up a significant proportion of the overall value-creation potential of conferences across all scales: individual, organisational and community (Edwards *et al.*, 2017). These benefits include knowledge gains, professional development, recognition and reputation, friendship, networking and collaboration (Jago & Deery, 2005; Severt *et al.*, 2007; Mair & Thompson, 2009; Foley *et al.*, 2010; Foley *et al.*, 2014; Mair & Frew, 2018; Hansen *et al.*, 2020). Exploring these less tangible benefits contributes to a broader understanding of the potential that business events hold for creating scalable and significant positive outcomes based on the exchange of knowledge and experiences between participants. The challenge for the business events industry, therefore, is to ensure skilled professionals have the information, skills and resources they need to enable them to create an event environment that maximises the likelihood that such exchanges, opportunities and, hence, benefits occur.

The following case study is an example of what Cohen and Levinthal (1990) refer to as the absorptive capacity of individuals and organisations to maximise the process of knowledge sharing and gaining during and after conferences via knowledge spill-overs. These spill-overs then lead to what are termed 'spirals of knowledge' (Nonaka, 1991; Nonaka & Takeuchi, 1995), where shared knowledge transcends the short-lived experience of the event to impact wider communities. The case study demonstrates how effectively planned, organised and managed conferences can provide a setting deliberately designed to ensure these important knowledge expansion activities take place.

## Case study: Knowledge sharing and organisational development through the conference EuroHeartCare 2017

| | |
|---|---|
| **Business event(s)** | EuroHeartCare (EHC) 2017 conference |
| **Location(s)** | Jönköping, Sweden |
| **Type of event(s)** | Major healthcare conference |
| **Event purpose** | To present and share the latest findings on cardiovascular heart care |
| **Primary audience** | More than 360 delegates made up of PhD students, nurses, university professors and researchers involved in cardiovascular heart care |
| **Primary event legacy** | Knowledge gains for delegates attending the conference and subsequent knowledge sharing with peers at their workplaces, resulting in better patient outcomes |
| **Additional legacies (beyond tourism)** | Multiplier impacts of knowledge gain via knowledge sharing with non-attending peers and sponsoring institutions (i.e. hospitals) which further improves patient outcomes, development of professional networks and collaborations, and student learning opportunities |
| **Summary quote** | "(Euro) is one of the most important platforms for driving research publications, exchanging ideas and forming deepening collaborations. Attendees can take knowledge back to their institutions (hospitals or universities), to develop practices and share ideas." |

This case study was adapted from: Gambina, A., Hagström, J. and Petzold, A. (2018)

### Introduction

The following case study looks at the 2017 EuroHeartCare (EHC) conference, which is widely regarded as one of the major drivers of research, ideas and collaboration in cardiac health care. Delegates who attend the conference come from a wide range of backgrounds, which creates opportunities for interactions across a range of health-care industry sectors.

Organisers actively encourage delegates to network and share their experiences in order to maximise collaboration and learning opportunities. In addition, delegates sent to the conference by the institutions in which they are employed (community health centres, hospitals, research institutes etc.) are strongly encouraged to share what they have learned with their peers upon returning to their workplace. In this way, the knowledge generated, shared and accumulated at the conference can be seen to stimulate a multiplier effect, where knowledge is shared and put into practice in multiple ways in order to achieve multiple legacies. This is not dissimilar to the way tourism spend, as a result of conferences, is said to create multiple economic benefits for host destinations (in simple terms, how many times the money spent by a tourist circulates through a country's economy).

The case study also highlights how the event provides professional development and recognition for practitioners, and offers researchers and post-graduate students opportunities to interact with those working in the field (including academic journal editors), in order to inform, focus or redirect their research and publications.

**Photo 5.1:** 2017 EuroHeartCare

## About EuroHeartCare 2017

EuroHeartCare is the annual meeting of the European Society of Cardiology (ESC) Council of Cardiovascular Nursing and Allied Professions (CCNAP). It aims to support healthcare professionals in delivering the best care possible to patients with cardiovascular disease. People from more than 40 countries around the globe were present at the 2017 conference, with most attendees coming from Europe. The event was hosted by Jönköping University and took place 18-20 May at the Culture Hall Spira in Jönköping. The EuroHeartCare Conference is one of the most important platforms for driving research publications, exchanging ideas, and forming or deepening collaborations within the field of cardiovascular nursing in Europe.

# Theoretical background

Meetings, conferences and business events strengthen communities with a joint interest in generating and sharing ideas through engagement in collaborative connections and business partnerships (Wenger & Snyder, 2000). However, organising and hosting such events requires considerable effort and allocation of often limited human and financial resources. This prompts some universities, hospital departments and cities to question the merits of such an undertaking. However, when potential hosts are clear on what they want to achieve from a conference, and when legacy outcomes are planned for at the outset, it is likely the benefits generated will far outweigh the costs.

Several studies have identified, documented and measured the economic impacts of conferences (Kim *et al.*, 2003; Hanly, 2012; Jones & Li, 2015). These studies often include economic multiplier effects that occur as a result of staging an event, whereby growth within a regional or national economy is stimulated via delegate spending. Whilst some of these studies have been criticised in relation to their applicability across different geographic locations or host city demographics (Hughes, 1988; Lee & Back, 2005), most have shown substantial direct economic impact on (at least) the local economy (Grado *et al.*, 1997; Kim *et al.*, 2003; Jones & Li, 2015).

In recent years, however, there has been growing interest in studies which document beneficial outcomes that are less-tangible, such as long-term impacts on industries, economies, professions, communities and knowledge. In particular, the Joint Meetings Industry Council (JMIC) has been involved in promotion of the "Beyond Tourism Benefits" (BTB) initiative, which offers ground-breaking research into a range of legacy outcomes beyond the economic tourism benefits of conferences (Edwards *et al.*, 2011). While direct economic effects are relatively easy to measure and communicate using traditional methods, these intangible benefits are more difficult to quantify. For example, there is not yet a standardised method for measuring how a conference has influenced a local research community, improved delegate and speaker careers. or enhanced the brand value of local organisers.

# Purpose of study

This case study on the European conference, EuroHeartCare 2017 in Jönköping, Sweden, seeks to contribute to the body of evidence in support of the JMIC initiative, by capturing and sharing case study data related to outcomes perceived by delegates to have occurred as a result of the conference.

The study gathers information on delegates' expectations prior to attending the conference, as well as opinion about the outcomes they perceived had occurred following it. There is also discussion on how closely the expectations and outcomes were aligned, and what was done by conference organisers to facilitate such alignment. The study was undertaken soon after the conference was held, hence most of the outcomes noted relate to short-term benefits.

## Methodology

Methodological approach: primary and secondary sources utilised; data obtained via face-to-face and online interviews, through observations at the conference and analysis of news/media releases.

Primary data were collected by conducting 15 face-to-face and online (Skype and email) semi-structured interviews with delegates, conference organisers and high-level decision makers at both Jönköping University and Jönköping municipality. These interviews were conducted prior to, during and after the conference. Observation of delegate behaviour at formal conference sessions and social events was conducted, and information was collected through news and media releases (Einebrant, 2017; EuroHeartCare, 2017; Manipal College of Pharmaceutical Sciences, 2017).

The data collected was analysed using an inductive approach (Thomas, 2006). All face-to-face and online interviews were recorded, and later transcribed. The text was analysed for key words and phrases, and a final set of common categories was developed. The findings were divided into two main sections: the attendance motivation felt by delegates *prior to* and *during* the conference (why delegates were attending); and second, the *post-conference* outcomes that were perceived by various stakeholders (delegates, speakers and organisers) to have occurred.

A number of factors impacted the research process and potentially limited the extent to which findings could be generalised. First, the interviews that were conducted prior to, and then after the conference were not carried out by the same set of people, hence comparisons between expectations and outcomes from EHC are less precise. This was because there was a very short time period allowed for data collection and some people interviewed in the pre-conference phase were not available post-conference. This is an example of why it is important that the research process is built into event planning at the outset.

Second, the case study makes use of interviews as the only source of primary data. Future studies could expand on primary data collection to include quantitative measures, given the appropriate tools and resources.

Finally, the research was conducted soon after the completion of the conference, hence long-term outcomes (2–5 years post-event) are not included. This makes it difficult to measure legacies in terms of organisational effectiveness or patient outcomes.

## Findings

Based on interview responses, the following findings are presented. Delegate motivations for attending EHC 2017 prior to the event are identified and then post-conference perceptions of outcomes (of both delegates and organisers) are discussed.

## Prior to conference: Motivations for delegates

Prior to the conference, the reasons delegates gave for wanting to attend were having opportunities to gain and share knowledge, develop professionally, network and achieve personal recognition.

### *Knowledge sharing*

Knowledge sharing is when individuals mutually exchange their knowledge and jointly create new knowledge (Van den Hooff & De Ridder, 2004). This means that knowledge sharing consists of both giving and receiving knowledge (de Vries *et al.*, 2006) and is a two-way process of knowledge formation between the sender and receiver of information. This is a common motivation for delegates at conferences (Adler & Kwon, 2002).

Interview data revealed most delegates attended the conference with the intention of sharing the knowledge gained with their colleagues back at the workplace. However, this was not the only type of knowledge sharing that delegates expected to take part in. They said they intended to take time to discuss common areas of interest, both in formal and informal settings. This kind of interaction could have various consequences, some of which are discussed later. Clearly, however, delegates' main goal was to learn from each other, as detailed in the following comments.

> *"Benefits of attending the conference include the knowledge you get together, and get to share with colleagues, which means everyone in the health center."*
> (Lena: delegate and nurse)

> *"The knowledge that the attendees can take back to their institutions (hospitals for clinical nurses and neuro specialists), and also of course we hope that people take knowledge back to the universities, perhaps to a lesser extent to develop practices, and share ideas, and to build on known knowledge."*
> (Tina: presenter at EHC 2017 and CCNAP member).

### *Knowledge gain*

When knowledge is passed on from one person to another without the receiver contributing towards the exchange, then the receiver is gaining knowledge within a specific field or subject. This one-way communication knowledge gain has been identified as an important reason for delegates to attend conferences. Many delegates like to simply sit and listen to speakers, and to take part in various scientific programs (symposia and workshops). By doing this, delegates can improve their skills and understanding of the subject being discussed via knowledge gain, which was another motivator for some delegates.

> *"I will listen to workshops and pay special attention to the English medical terms used by the speakers, which is very important to me right now when writing papers"*
> (Louise: delegate and PhD student).

Furthermore, Louise claimed that one of the advantages of attending the conference was having the latest information all in one place for certain topics in the healthcare sector:

5

*"We get to hear about the latest research, gathered in one place, in three days."*

This helps delegates stay up-to-date with the latest revelations in the healthcare sector, where many speakers highlighted and debated what they thought was important.

*"My view is that conferences are important because you get to know the latest findings on certain topics. I think it would be awful if patients had to ask me a question and I wouldn't know how to answer. The patients could ask about the latest findings in atrial fibrillation, and if they had to do so, I would like to be up to date and give them information about it"* (Lena: delegate and nurse).

## Professional development

The goal for delegates when sharing and gaining knowledge is to enhance their skills and become better professionals at their place of work. The EuroHeartCare conference provided many opportunities for delegates to get involved and learn something new. The various symposia and workshops assigned a practical element to the conference, where nurses and other healthcare professionals were able to simulate real-life scenarios when treating patients.

*"I think other individuals here also feel that one chooses what one is interested in listening to, which might be of benefit to what one does due to what one works with"*
(Petra: delegate and nurse).

The conference was also well suited to students and researchers who believed they would be able to generate ideas for new potential papers, particularly as a result of the eight abstract-based sessions organised (EuroHeartCare, 2017).

## Networking

When delegates who share common interests are gathered within the same venue, opportunities arise to maintain and create new contacts. There were delegates that formally arranged to meet up with each other at a certain time and place, whilst others met up during an event at the conference and befriended each other as the conference progressed. It is important to note that both formal and informal interactions were important for delegates, and many noted that creating such networks was a strong motivator for attending.

Formal interactions usually entailed a pre-determined agenda, which both parties had agreed to discuss, whilst informal interactions were often held during breaks between workshops, or at a restaurant after the day's activities. Informal interactions provided the opportunity to discuss personal matters, which typically raised the probability of future collaboration (Jeong *et al.*, 2014).

*"We have a lot of discussions during the breaks, both formal and informal, and I think a lot of knowledge exchanges happen during these discussions. This conference allows you the time to sit down in a relaxed setting and just chat with other delegates. We have*

*the informal events such as the opening ceremony and cocktail party, which is a good opportunity for the delegates to discuss their research in a relaxed setting"* (Tina: presenter at EHC 2017 and CCNAP member).

*"I will also get the opportunity to talk to the attendees and exhibitors about my research area in English. Building networks in my field is something that I aim to do whilst talking about my research area"* (Louise: delegate and PhD student).

*"The EuroHeartCare conference provides me with an opportunity to create new international connections and collaborate with a wider range of people from diverse background with professional expertise in cardiology"* (Sangeetha: delegate and speaker).

### *Personal recognition*

Another reason for delegates to attend this conference was the personal recognition that they would be gaining. This was especially the case for delegates who were speaking at one of the events, discussion leaders, chairpersons, judges of scientific abstracts, or any other role that involved a certain degree of authority. For example, Sangeetha presented a paper on '*Assessing the effect of gender in the management of atrial fibrillation*' and was recognised for her efforts by receiving the 'Top Scored Poster' at the end of the congress.

*"Having bagged the Top Scored Poster award was indeed an honour for my entire team. This appreciation has not just boosted me to work even harder but has also been an impetus for my peer mates working towards a quality research affecting patient population"* (Sangeetha: speaker and PhD student).

Moreover, delegates attending the event got the chance to discuss their ideas and personal projects, which is another form of recognition.

## Prior to conference: Delegate motivations by attendee type

Generally, there were four types of attendees at the conference. They were:

- ☐ PhD students
- ☐ Nurses
- ☐ Senior professors
- ☐ Local (Swedish) delegates.

It was found that, within each group, there were similar motivations for attending.

### *PhD students*

PhD students had the opportunity to acquire knowledge and skills needed to write scientific articles. Editors of both American and European cardiovascular nursing journals were present at the conference. They took the time to discuss certain topics of interest in workshops and seminars. This meant that PhD students had the opportunity to listen to the latest findings in a medical area that could clarify their own research contribution.

5

*"We will have a workshop on how to write a scientific article, because quite a lot of participants will be PhD students working within this discipline. So, we will also have one workshop, where we will have editors from American and European cardiovascular nursing journals and they will act as mentors, where students can discuss what is important when writing or publishing a scientific article,"*
(Jan: local organiser and senior professor).

## *Nurses*

New guidelines, which focus on how a nurse should treat their patients, and new treatments, such as new ways of using existing equipment or new equipment being introduced, presented nurses with opportunities to learn new skills and gain knowledge. Selected speakers and event discussants shared their knowledge with nurses through presentations and lectures. Cardiopulmonary resuscitations (CPR) and electrocardiograms (ECGs) were two of the many workshops where nurses had the chance to practise and enhance their skills.

*"We will invite people from all over the world to share their competence with nurses. Attendees will also have the opportunity to gain new knowledge and skills from the presentations, lectures and workshops organised. When it comes to the practice of skills, nurses will have workshops on CPR (cardiopulmonary resuscitation) and ECGs (electrocardiograms)"* (Jan: local organiser and senior professor).

## *Senior professors*

Senior professors knew they would be meeting with other professors at EHC 2017. Therefore, the conference provided them with a chance to network amongst themselves.

*"From my own experience at many similar conferences such as EHC 2017, I realised that one of the most important features of a conference is the opportunity to meet and discuss with other researchers in the same field of study. I am looking forward to meet (and have already planned some meetings) with colleagues from other universities, both national and international,"* (Jan: local organiser and senior professor).

## *Local (Swedish) delegates*

Proximity to the conference, which was held in Jönköping, Sweden, meant it was easier and more convenient for Swedish delegates to be present at the conference. When Louise, a PhD student and Swedish delegate, was asked what attracted her to EHC 2017 she replied, "*The proximity to begin with, but more importantly we hear about the latest research, gathered in one place in three days.*"

*Jan, who was mentioned above, also viewed proximity as a major draw card, "It is a great opportunity for those working in this region to attend a conference that is quite close to home with no big distances of travel required".*

## Post conference: delegate perceptions of benefits

The beneficial outcomes delegates believed were generated as a result of the conference covered specific projects, professional development, knowledge gains, knowledge sharing, and networking opportunities.

### *Formation of an ongoing project*

A key outcome from EHC 2017 was the setting up of an ongoing project, where the main aim was to identify patients with atrial fibrillation (irregular heartbeats). The conference gave further information on patients living with this condition as a reminder of what had been learned. Lena (nurse) and her colleagues set up a one-day event, to which they invited nurses from 'Bra Liv Health Centres' (a group of health centres within the Jönköping region) to inform these nurses about the condition (atrial fibrillation) and to reinforce the new information acquired at the conference.

> *"On the 9th of November we will invite nurses from BRA LIV healthcare centres and pass on the knowledge we learnt at the conference about atrial fibrillation"* (Lena: delegate and nurse).

### *Professional development*

From the post-event interviews, career enhancement was widely cited as a benefit gained by delegates as a result of attending the conference. Through the various workshops and events, delegates felt they would be able to apply what they had learnt at the conference to their place of work. Guidelines presented at the conference that aimed to assist nurses in treating patients were considered useful. For researchers, the diverse backgrounds of healthcare professionals present at the conference assisted expansion of their knowledge within certain areas of research.

> *"The EuroHeartCare 2017 conference was the best way to connect with scientists working in different spectrums of cardiovascular sciences and this has helped me widen my horizon of clinical research and practice"* (Sangeetha: speaker and PhD student).

> *"The EHC conference has strengthened my professional capabilities and made me aware of not stagnating with old findings and old ways of working, but to rather continuously improve and develop myself as a cardiac nurse"* (Petra: nurse and delegate).

> *"I have been trying to work in line with the guidelines and successful research that was presented at the conference. The main aim of this is to better patient treatment"* (Petra: nurse and delegate).

### *Knowledge gains*

The conference provided delegates with the latest research within the cardiovascular field, which proved to be of importance to nurses and academics alike. Both types of professionals claimed they felt more up-to-date with regard to cardiovascular healthcare as a result of listening to the speakers and participating in the conference.

*"The symposia delivered by several excellent scientists and researchers have demonstrated the need for adopting standard methodologies for high quality research and potential interventions to be implemented by non-physician healthcare providers to improve cardiovascular care"* (Sangeetha: speaker and PhD student).

*"I feel that I have gained something from the conference, particularly what topics are being discussed in Europe on cardiovascular healthcare, especially due to the international profile of the delegates at the event"* (Lena: delegate and nurse).

## Knowledge sharing

Knowledge sharing was identified as a clear outcome of EHC 2017. It was stated in interviews that knowledge was shared both during the conference (with other delegates) and after the conference (with peers in delegates' workplaces). This meant that organisations could leverage the multiplier effect of knowledge gain against the cost of sending just one (or a few) delegates to the conference.

*"Whenever attending a conference, the people present have always shared what they learnt with their colleagues back at the health care center"* (Lena: delegate and nurse).

*"With a background in clinical pharmacy and research, and special interests in preventive cardiology, I have had the opportunity to interact with several delegates who shared the common interests and it was quite inspiring to learn about their contribution in this field"* (Sangeetha: speaker and PhD student).

*"I have had the possibility to present what was learnt at the conference with my colleagues at the health center by creating a PowerPoint presentation with what I thought were the main takeaways from the event"* (Petra: delegate and nurse).

## Networking

EHC 2017 was a great opportunity for people who share common interests to meet and discuss their projects and form collaborations. It was noted that forming collaborations and mingling with people at the event really depend on an individual's openness to take a chance and present themselves and their ideas to other people.

*"As a young researcher, I have been very fortunate to connect with some of the best experts in the area who not only helped me understand my potentials and possible career options to explore but also turned out to be great mentors who still continue to guide me with my research and publications. Several interesting discussions have also paved way to collaborate with other institutions and for a multicentric study,"*
(Sangeetha: speaker and PhD student).

*"It's up to the delegates to take the opportunity and not only listen to the presentations but also discuss with people who are present at the conference,"*
(Jan: local organiser and senior professor).

## Other beneficiaries from the conference

### *Jönköping University (JU)*

The EHC conference venue delivered a range of benefits to the university, its employees and its students.

PhD students at the university had the opportunity to meet both foreign and local professors. This meant that they could make contact with professors who could possibly help them in their field of work. Contacts with foreign professors also made it easier for local PhD students to go on exchange, if interested, as they had a reference point when applying to the university in question.

> *"I think that one of the greatest opportunities for PHD students during EHC 2017, was to get the chance to meet researchers that you have read about in your studies ... also if a student would like to go on an international exchange program, it would be much easier to do so if the student has a contact at a university abroad"*
> (Jan: local organiser and senior professor).

There were also workshops, such as 'Publishing your articles with impact and influence', which were aimed at identifying the main tools and skills needed in order for PhD students to publish high-quality articles in journals. These workshops could assist PhD students' publishing endeavours.

> *"We had a workshop called 'publishing your articles with impact and influence', which was mainly directed towards PHD students, and those who attended that workshop had of course great interest in getting more knowledge on what makes a good article"*
> (Jan: local organiser and senior professor).

### *Healthcare centres*

Of all delegates at the conference, 54% were nurse practitioners (EHC, 2017). These nurses participated in the various events at the conference, gained knowledge and discussed ideas with other delegates before returning to their respective healthcare centres or workplaces (including hospitals and clinics). Petra and Lena (both nurses) stated that the knowledge gained from the conference was shared with their colleagues through various channels. This means the benefits of attending the conference were passed on to other nurses at the health centre who did not attend the conference. Hence, the conference contributed to an improved environment at the healthcare centres where the delegates worked.

> *"When one attends a conference and shares knowledge, the people at the healthcare center get onboard which creates a feeling of togetherness"* (Petra: delegate and nurse).

### *CESAR Group*

The CESAR group is a group of researchers who work toward collaboration and exchange in Swedish cardiovascular academic research. This group consists of researchers from different universities across Sweden, mainly from Jönköping, Kalmar, Linköping and Nörrköping. EHC 2017 was an opportunity for the CESAR group to collaborate and exchange information on cardiovascular research.

> *"It would have been impossible to have this conference in Jönköping without the help from CESAR group. We have been a well-functioning research group for many years and have published a lot of research together. This conference was a perfect opportunity to manifest that. We had a CESAR meeting just before the conference started and during the meeting, we had a lot of discussions, both within the group, but also with other research networks. One example is the three-day writing seminar we had at Linköping University that took place after the conference with a research group from University of Kentucky, US"* (Jan: local organiser and senior professor).

### *The destination research community*

Scandinavia is known to be research-orientated when it comes to cardiovascular health care. The conference reflected the research that was being undertaken within the area, as many high-quality researchers from all over Sweden attended. This, in turn, strengthened the destination researchers' level of expertise.

> *"The program that we've developed here is more along the lines of research, as delegates in Sweden are more interested in this line of work. Whereas two years ago, when we were in Croatia, Southern Europe, the program was very much about the basic elements of nursing and similar areas. So the program is adapted according to the country we host it in"* (Damian: EHC 2017 organiser and CCNAP member).

## Discussion of findings

An important aspect of this case study was recognising the multiple levels of knowledge gain and sharing generated as a result of the EHC 2017 conference. From interview data, it was apparent there was at least a three-stage knowledge transmission process (Figure 5.1). It is important to note that, for this study, the outcomes realised in Stage 3 are only short-term because of the relatively brief period which has elapsed since the conference took place. It is anticipated that more organisational benefits would have been evident if the time frame for data collection and analysis had been extended. This should be an objective of future studies seeking to make a valuable contribution to the body of meetings and business event literature.

### Comparing expectations and outcomes

Common expectations and outcomes for delegates were both knowledge sharing and knowledge gaining. Delegates are usually sent to conferences on behalf of organisations in order for them to learn something new and bring the information back to the workplace. The knowledge brought back is expected to be shared with colleagues who work on the same or similar projects.

Figure 5.1 represents the process of knowledge transmission from the delegates at the conference to the people at their respective organisations.

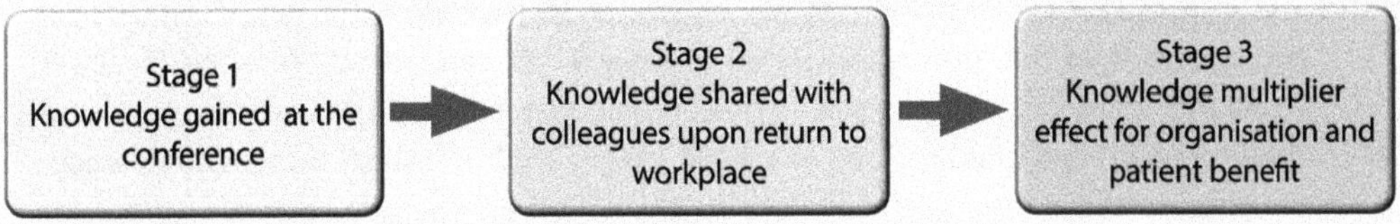


**Figure 5.1:** The process of knowledge transmission

## *Stage 1 – Knowledge gained at the conference*

Delegates gained knowledge at the conference via symposia, program abstracts, practical workshops, four special events (inaugural session, networking reception, CCNAP general assembly and the closing ceremony), keynote speeches, and discussions with other delegates (EHC, 2017). The knowledge gained occurred both during one-way communication events (where delegates usually sat down to listen), and two-way communication exchanges (such as formal and informal conversations between delegates outside formal sessions).

## *Stage 2 – Knowledge shared with colleagues*

After the conference, delegates returned to their respective organisations and shared what they had learnt with their colleagues through presentations and scheduled meetings. For example, Lena and Petra (both nurses) compiled a short document highlighting key takeaways from the conference and presented a summary of it at a workplace meeting.

## *Stage 3 – Outcomes for organisations*

The last stage in this process results in the organisation being better informed in specific areas related to topics discussed at the conference. This could lead to various outcomes, such as new equipment purchases, more effective patient treatment practices, formation of a project (as discussed earlier) or a rise in interest, and increased academic writing, in areas related to cardiovascular healthcare. Interview data in this case study reflected outcomes related to the formation of a knowledge-sharing project and increased activity/interest in research and publications related to cardiovascular healthcare.

> *"We know that before and during the conference, a lot of writing is going on since students and professors need to write an abstract and send it in (for the conference). In fact, a lot of research groups gather and discuss what kind of research is ready to publish, and how scientific papers can be collected within a common framework, so we always have a little surge in paper publication after a conference"* (Jan: local organiser and senior professor).

## Conclusions

This study makes a contribution toward a better understanding of the range of beneficial legacy outcomes that can be generated as a result of a conference. More specifically, it highlights some of the important non-economic benefits of conferences by analysing the short-term outcomes realised as a result of EHC 2017. The study focuses on delegate benefits realised at the conference, but also discusses the consequent benefits realised for the organisations in which they work and the patients for whom they care.

From the interviews carried out, it was clear that knowledge gain and knowledge sharing were two important benefits that made the EHC 2017 conference worthwhile. For delegates (predominantly nurses and university professionals), gaining knowledge in the healthcare sector meant they felt more competent in their profession, with many stating they anticipated the conference would positively and significantly impact their professional development. Those interviewees said employees at organisations such as healthcare centres want to attend conferences in order to hear about the latest treatment options and relevant research, and later share this knowledge with their colleagues.

# Theoretical discussion

Conference attendees building on, retaining and passing on knowledge constitutes what is termed "absorptive capacity" (Cohen & Levinthal, 1990). Absorptive capacity is connected with generating knowledge spillovers and the progression of spirals of knowledge, resulting in impacts in a broader context. This section explores absorptive capacity and its relationship to the ensuing spirals of knowledge which may be generated as a result of business events.

## Absorptive capacity

Van den Hooff and Ridder (2004, p.118) state knowledge sharing implies that at the one end of the exchange there is someone "donating knowledge" (communicating intellectual capital) and at the other end, someone "collecting knowledge" (acquiring others' intellectual capital). Gaining or acquiring knowledge refers to the capacity to store new knowledge in the memory, to later remember it and put it to use (Cohen & Levinthal, 1990).

This capacity to collect or acquire knowledge is intrinsically linked to what Cohen and Levinthal (1990) defined as the "absorptive capacity" of individuals and organisations to improve creativity and innovation. Originating in firm theory (Caragliu & Nijkamp, 2012), it considers a firm's ability to absorb (identify, assimilate or transform) and then apply external knowledge, and suggests that organisations (and individuals) have a limit to the amount of information they can absorb (or the efficiency with which they can absorb it). Over time, the concept has been applied in other fields, particularly those dealing with the processes of knowledge acquisition, information dissemination and innovation enhancement (Caragliu & Nijkamp, 2012).

Under this construct, the ability to internalise and use external knowledge depends on the existence of some level of prior related knowledge. According to Cohen and Levinthal (1990), preceding knowledge "includes basic skills or even a shared knowledge but may also include knowledge of the most recent scientific or technological developments in a given field" (p. 128). Thus, absorptive capacity is understood as the process by which individuals and organisations are able to recognise the value of new information, assimilate it and apply it (Easterby-Smith *et al.*, 2008). This requires them to build on prior knowledge (including past experiences, culture and information) and on their retention capabilities. Furthermore, the absorptive capacity of an organisation will depend upon the absorptive capacities of its individual members and on their ability to use that knowledge. Thus, Cohen and Levinthal claim that the capacity (systems, processes, attitudes, personnel resources) for internal transference of acquired knowledge across organisational levels and units is just as important to absorptive capacity as exposure to external knowledge itself.

Critical to this construct is that absorptive capacity facilitates processes of knowledge sharing and gaining (Kwok & Gao, 2005; Van Wijk *et al.*, 2008; Donate & Guadamillas, 2010). The absorptive capacity of individuals has been associated with their "openness to new ideas, eagerness to learn, and innovativeness in approach" (Dalkir, 2005, p. 212). Hence, how knowledge is absorbed depends not only on the predeterminate knowledge base (e.g. education and professional background), but also on the dynamic capabilities of knowledge acquisition and assimilation (*potential capability*), and transformation and utilisation (*realised capability*) (Zahra & George, 2002). The dynamic dimensions of absorptive capacity proposed by Zahra and George (2002) also suggest that a recipient's predisposition and readiness to gain new knowledge impacts the efficacy of knowledge transference. Accordingly, the higher the absorptive capacity, the better the conditions for sharing knowledge and gaining new knowledge (Chung-Jen, 2004).

Given these theoretical contexts, we can begin to understand why conferences are optimal sites for facilitating acquisition, generation and transfer of knowledge. Efficiency in knowledge sharing and gaining as a result of attending conferences can be related to two key factors: the familiarity of attendees with the existing knowledge base, topic or subject, and the personal expectations, motivations and readiness to learn which delegates hold. Delegates attend conferences for many reasons; however, the vast majority are there to learn, develop, connect or share. Knowing they will be gathering with like-minded peers who share those same goals is often motivating and inspiring, and successful conference environments create a buzz, a space and a framework where such goals may be achieved. These factors can have a significant influence on delegates' level of engagement and their ability to absorb the knowledge being exchanged (Gupta & Govindarjan, 2000).

With this in mind, it is useful to consider the 4E framework of training methods. This framework describes the varying intensities of learning one may experience when being exposed to new knowledge. The framework identifies four important strategies which trainers can use to facilitate learning; *exposure, experience, expertise* and *embedding* (as proposed by Farkas *et al.*, 2003). This framework can also be drawn upon to explain how knowledge transmission, dissemination and utilisation might be facilitated via conferences.

## Exposure

*Exposure* strategies use creative and innovative methods to help learners acquire, assimilate and use new knowledge. Previous literature suggests knowledge dissemination should position learners as active participants and information-seekers rather than as passive recipients of information (Farkas *et al.*, 2011). Such strategies help learners explore, consider and confirm information. They seek to increase the interactive nature of communication in learning, where participants are encouraged to be actively involved in the learning, rather than passively reading or listening to information. They are the least intense (and usually first) level of training (learning) and include things like lectures, workshops, webinars and discussions, all of which are integral components of business events.

Individuals and organisations can develop their absorptive capacity (amplify the pre-existent knowledge base and improve potential capabilities) by learning in environments where motivation or engagement is heightened. Conferences and business events provide just such an environment, where the shared base of existing information, the instantaneous accessibility of new information, and the gathering of like-minded peers and experts in one's field significantly heightens one's motivation and engagement and makes information easier to absorb.

## Experience

*Experience* strategies also deliver new information, but here there is additional emphasis on creating positive *attitudes* in the learner towards not only the *acquisition* of external knowledge but also the motivation to *use* it. Examples of this include:

- Mentoring (i.e. where researchers inspire or mentor students to discover their own areas of future study);
- Observing demonstrations (including observing new equipment and/or methods);
- Participating in live events (including podcasts, webinars etc.) incorporating exercises and discussions that relate to deeper understanding of information or values;
- Meeting peers who can clarify or inspire

(Centre for Psychiatric Rehabilitation, 2021).

Experience relates to the level of effort required to make associations between existing and new knowledge, and therefore, the capacity to acquire, assimilate and apply that knowledge. Again, these types of experiences are able to be delivered in a well-planned and targeted business event program. Conferences frequently involve mentoring of new researchers, demonstrations of how new knowledge can apply in real world settings, and meetings of peers and colleagues which lead to new ideas and understandings.

## Expertise

*Expertise* strategies seek to increase competency in a skill or practice. Examples might include things like internships or supervised practicums. The assimilation or integration of pre-existing and new knowledge also depends on the richness of the former. As Cohen and Levinthal (1990: 131) argue, "learning is cumulative, and learning performance is greatest when the object of learning is related to what is already known". Consequently, the more diverse knowledge an individual possesses, the more capacity they have to assimilate new information by relating it to the scope of what is already known. Conferences are an ideal context in which a broad range of expertise and knowledge can be shared, demonstrated, and gained through presentation and discussion of the latest research. This case study also demonstrates that conferences which implement strategies that specifically target sharing of knowledge and practices **prior to, during and post the event** can directly lead to delegates' heightened capacity to learn new techniques in order to increase the effectiveness of their practices. Additionally, the case study exemplifies the improvements in organisational processes and systems that can be achieved as a result of sharing knowledge, guidelines and best practice following an event.



## Embedding

*Embedding* strategies refer to methods which help an organisation (or individual) internalise or embed new information into its (their) ongoing practice. Once information is assimilated, the organisation (or individual) is able to apply those new associations and linkages in novel and creative ways (Edwards *et al.*, 2017), generating an environment for innovation and utilisation of knowledge in more diverse contexts. This relates to the case study in a number of ways.

First, one of the most effective ways for individuals to embed or assimilate new knowledge is to be required to teach that knowledge to others. Whilst the process of teaching may be different to actual learning, teaching requires that one thinks deeply about the concepts to be presented, considers alternative perspectives and is able to respond to a range of questions or challenges that may be posed in relation to what they are teaching.

Second, many organisations find themselves exposed to new knowledge, but unless that knowledge is institutionalised (or internalised) into daily practice it is likely to be lost. For example, when an organisation relies on the skills of certain key staff members without documenting the processes they go through when performing their job, they risk losing that knowledge when the staff member moves on.

By encouraging delegates to 'teach' others at their workplace what they learned at a conference, and by requiring them to document the information they acquired, new knowledge is embedded in the organisation's resources and also in the minds of the delegate representatives. In this way, the case study demonstrates how the full benefits of knowledge creation and spirals can be obtained by the delegates involved, the organisations they work for, their field of research and the wider community. Conferences also provide opportunities for direct feedback that leads to further innovations in the daily practices of attendees.

## Conferences as drivers of spirals of knowledge

Conferences provide temporary contexts for learning and interaction, triggering intense knowledge exchange mechanisms (Maskell *et al.*, 2006) within a particular subject, discipline or industry sector (Edwards *et al.*, 2017). These events operate as nodes or "temporary clusters" (Maskell *et al.*, 2006) that create linkages between people of diverse spatial scales (local, regional, national and international). They are an efficient way of transferring knowledge about the state-of-the-art on a topic or problem, new developments or ideas, future directions, and to collectively discuss the relevance of these issues (Bathelt & Henn, 2014).

As a result of these discussions and exchanges of ideas, conferences become hubs for innovation and changes in social practice (Edwards *et al.*, 2017). Besides enabling knowledge flows, conferences set the stage for knowledge creation as a result of the emergent collaborations and networks between attendees. As stated by Edwards *et al.* (2016: 2), "Knowledge creation is increasingly dependent on collaboration, which relies on the camaraderie, trust and relationships formed through face-to-face contact. Business events provide the platforms for this physical interaction and the foundation for the networks that then drive collaboration and innovation."

Knowledge is not an easy concept to define owing to the complex nature and interrelation of its constitutive elements. In this regard, Davenport and Prusak (1998: 5) observe that knowledge "is a mixture of various elements; it is fluid as well as formally structured; it is intuitive and therefore hard to capture in words or understand completely in logical terms". Although knowledge is considered to be an individual asset held in people's minds (Smith, 2001), knowledge creation is essentially a social process (Nonaka, 1991, 1994; Nonaka & Takeuchi, 1995). Moreover, knowledge is continuously generated, reproduced and transformed through relational processes based on communication and actions between humans (Stacey, 2001; Chua, 2002). As Stacey (2001: 4) claims, "knowledge is not a 'thing' or a system, but an ephemeral, active process of relating."

Despite the flourishing of knowledge management theories, little attention was paid to knowledge itself and how knowledge was created until the 1990s (Nonaka & Takeuchi, 1995). Such theories were mostly concerned with issues of control and exploitation of existing knowledge, rather than seeking to comprehend the formation of new knowledge within organisations and beyond organisational boundaries. In addressing this theoretical vacuum, Nonaka (1991, 1994) and Nonaka and Takeuchi (1995) proposed a theory of organisational knowledge creation that explain how the interaction between processes of knowledge conversion produces continuous spirals of knowledge that lead to problem-solving and innovation.

This theory is premised on the idea that knowledge is created in context-specific situations where information is interpreted by individuals based on their value systems. Therefore, instead of conceiving knowledge as in the western (static and context-free) tradition of 'justified true belief', the authors emphasise its dynamic and humanistic nature, defining it as "a dynamic human process of justifying personal belief toward the 'truth'" (Nonaka & Takeuchi, 1995, p. 58). Embedded in this definition are the two fundamental dimensions in knowledge creation (Nonaka 1994, p. 15):

1. The *epistemological* dimension – the role of interaction in developing ideas that are held in people's minds, and therefore, the importance of "communities of interaction" in contributing to amplifying and developing new knowledge
2. The *ontological* dimension – the exchange of knowledge across boundaries (intra- and inter-organisational) in communities of interaction which define "a further dimension to organisational knowledge creation, which is associated with the extent of social interaction between individuals that share and develop knowledge.

The situational embeddedness of knowledge is defined as '*Ba*', a Japanese philosophical concept that Nonaka and colleagues (2001, p. 19) describe as the "shared context in which knowledge is shared, created and utilised". Ba is the time-space that attributes meaning to the information that individuals in their interactions transform into knowledge. Conferences are such time-space platforms for physical interactions. Moreover, Ba exceeds the physical by bringing continuity to the time-space nexus through the interpersonal relationships created in the form of networks, collaborations and other kinds of communities of interaction. In other words, we can think of conferences and their legacies as Ba; as the context 'in motion' that carries meaning and facilitates the creation of new knowledge (Nonaka & Toyama, 2003).

A critical contribution of this theory lies in the role attributed to tacit and explicit knowledge and the modes of conversion of one into the other that results in new knowledge. Michael Polanyi (1966) first introduced the differentiation of human knowledge between the explicit and the tacit, to foreground the validity of the latter and its contribution to comprehension and discovery, comprising all the things we know but we can't tell. Drawing on Polanyi's work, Nonaka and Takeuchi (1995, p. 59) define explicit (or codified) knowledge as "transmittable in formal, systematic language", while tacit knowledge is "personal, context-specific and, therefore, hard to formalise and communicate". Explicit knowledge is, thus, objective, easy to process, transfer and store through verbal, mathematical and numerical language, condensed as in the *know-what* (e.g. data, scientific formulae, specifications, manuals) (Nonaka *et al.*, 2000; Stacey, 2001). On the other hand, tacit knowledge is subjective, deeply rooted in action, experience, values, ideas, ideals and emotions (Nonaka *et al.*, 2000). The territory of tacit knowledge is that which is not explicated (Collins, 2010). Therefore, it is associated with individuals' cognitive frames and mental models that manifest in their technical and cognitive skills, the *know-how* (routines, intuition, common sense).

Tacit knowledge has been largely addressed in knowledge management theories because it is regarded as a critical component in "leveraging the overall quality of knowledge" (Smith, 2001, p. 312). Nevertheless, it is worth noting that many of these theories have addressed both types of knowledge as opposites, when they are, in fact, inexorably intertwined (Tsoukas, 2005). As Collins (2010) argues, although it has been widely accepted that all explicit knowledge is based on tacit knowledge, we cannot grasp the tacit without the explicit. Consequently, in order to explain processes of knowledge creation, it is critical to understand the relationship between explicit and tacit as complementary or reciprocal (Nonaka *et al.*, 2001).

Based on the distinction between explicit and tacit forms of knowledge, and building on the notion of "knowledge conversion" proposed in the field of cognitive psychology by Anderson (1983), the authors identify four modes of knowledge conversion (Nonaka & Takeuchi, 1995; Nonaka & Toyama, 2003) as a result of the interaction amongst individuals within a specific context (see Figure 5.1):

1. Tacit to tacit: this conversion occurs through processes of *socialisation* in which individuals acquire knowledge by sharing experiences and perspectives, and thus it requires physical proximity and spending time together (e.g. observation, conversation, joint activities).
2. Tacit to explicit: this is the process of *externalisation* that is grounded on continuous and meaningful dialogue. Tacit knowledge becomes explicit by, for example, using metaphors, analogies and models to articulate and reveal in discourse that which is not codified in the mind. Mental models, perceptions, experiences, hunches and so on find, through metaphor, a way to be manifested. Metaphors are not merely a linguistic device; they are related to the human thought process (Lakoff & Johnson, 2008). When this mode of conversion occurs, knowledge is "crystallised", and so it is then possible to share it with others, becoming the basis for developing new knowledge.
3. Explicit to explicit: this knowledge conversion implies the *combination*, editing and processing of diverse sources of explicit knowledge (intra- and

inter-organisational), generating more complex bodies of explicit knowledge (e.g. new documents, databases).

4. Explicit to tacit: through experimentation and iteration of trial and error, explicit knowledge is converted into tacit by processes of *internalisation*. Usually referred to as "learning by doing", internalisation is the amalgamation of explicit (outside) knowledge and individuals' mental models, practices and skills.

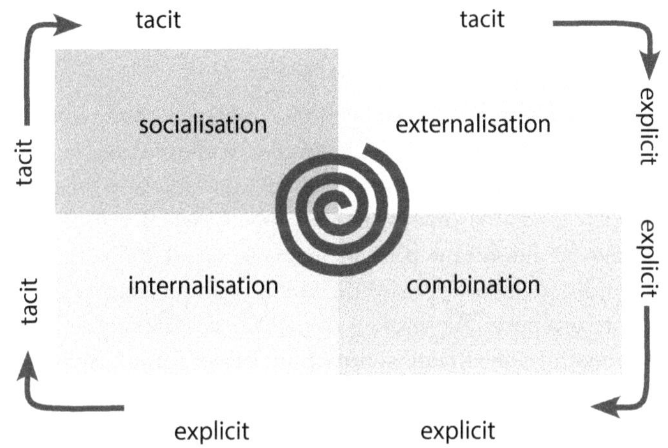


**Figure 5.1:** The SECI model (Nonaka & Takeuchi, 1995)

The 'SECI model', as it was coined, explains the social process of knowledge conversion, which "is not confined to within an individual" (Nonaka *et al.* 2001, p. 14) as it requires the exchange of knowledge held by different people. Therefore, this model is the result of a continuous dialogue between explicit and tacit knowledge at the micro (individual) and macro (intra and inter-organisational) levels. Consequently, both types of knowledge result in expanded quality and quantity (Nonaka, 1991; Nonaka & Takeuchi, 1995; Nonaka *et al.*, 2000).

Conferences, as privileged settings for knowledge exchange, are key drivers in promoting processes of knowledge conversion. Moreover, as the shared information is converted into knowledge during and after conferences take place, it contributes to the development of what Nonaka (1991) and Nonaka and Takeuchi (1995) defined as "spirals of knowledge". The constant movements between the different modes of knowledge conversion and across the different ontological levels (micro-macro) form a spiralling process, in which new interactions trigger the creation of new spirals of knowledge. The process begins at the micro-level with individual knowledge and moves upwards by constant conversions resulting from interactions, enriching knowledge and amplifying the scope of the spiral of knowledge. Knowledge creation is, therefore, a continuous process "from being to becoming", in which the individual not only transcends their own boundaries, but also organisational boundaries, as new knowledge outpours into local and global communities to create ongoing knowledge gain.

Thinking of conferences as privileged contexts for creating knowledge, Ba, in the way portrayed by Nonaka and Takeuchi, entails delivering conditions to prompt the different processes of knowledge conversion. To do so, it is worth taking into consideration Hoe's (2006) reinterpretation of the SECI model, narrowing it down into two main processes of knowledge creation: informal (rich in tacit knowledge, i.e. socialisation and internalisation), and structural (dense in explicit knowledge, i.e. externalisation and combination). Structural processes of knowledge creation are found in conferences because codified and formal language transferred in a systematic way is the conventional format of transference in presentations and keynotes.

However, the subjective (and often unconscious) nature of tacit knowledge means that it is more challenging to communicate in the standardised organisation of conferences. Thus, informal, dynamic and unique processes and activities (such as coffee breaks, shared meals, social events and other joint activities) play an important part in business event delivery because they provide the conditions and opportunities to encourage sharing of knowledge and bonding between peers. As Maskell *et al.* (2006, p. 1002) argue, informal environments can "sometimes be the most important places for information exchange in temporary clusters". Events which include planned, targeted strategies to maximise the potential such interactions hold can build stronger ties between attendees and increase tacit knowledge transfer (Augier & Vendelø, 1999; Hansen, 1999).

In sum, knowledge flows during and after conferences support knowledge acquisition and transformation and the emergence of new knowledge. Put simply, knowledge gaining and sharing are principal motivations for most business event attendees, or as Oestes *et al.* (2017) stated, "this is why conferences matter". Those industry professionals who are able to design and present business events which significantly increase the generation, sharing and spirals of new knowledge will create legacies of far greater benefit than simply tourism expenditure.

## Lessons learned

Governing bodies, institutions, convention bureaux, peak organisations and private enterprise all invest in business events for a range of reasons. Individual delegates and the organisations that commonly fund their employees' attendance at business events also make a significant investment in terms of time, cost and inconvenience. It is important, therefore, that those responsible for the organisation, delivery and evaluation of these events ensure they are maximising the return on investment for all stakeholders. Whilst direct economic benefits have long been the most visible and measurable return identified, the industry is now able to consider other, less tangible, but arguably, more important returns. One of the most important of these for the well-being of society as a whole is the generation, acquisition and transfer of knowledge.

Two significant theoretical contributions to understanding how these processes occur are the absorptive capacity construct (Cohen & Levinthal, 1990) and the SECI model (Nonaka & Takeuchi, 1995). Cohen and Levinthal's absorptive capacity construct explains how the ability of individuals and organisations to assimilate and use external knowledge can contribute to the formation of knowledge spirals leading to innovation. Similarly, Nonaka and Takeuchi's model describes how different types of knowledge are transformed in context-specific situations creating virtuous spirals of knowledge. Their theory sheds light on an understanding of conferences as communities of interaction which lead to the development of the ideas that are held in people's minds and which contribute to the amplification and development of new knowledge. Importantly, these communities of interaction enable the exchange of knowledge across intra- and inter-organisational boundaries, various fields of study and geographic distances.

From Nonaka and Takeuchi we can understand how conferences have the capacity to epitomise the spiral of knowledge as they facilitate a constant movement of knowledge in which new interactions trigger the creation of new spirals. When we consider conferences at the micro-level, we can see how knowledge stored within an individual moves upwards as it is shared via constant conversions between attendees resulting from their interactions during and post conference. This continuously amplifies the scope of the spiral of knowledge.



Farkas *et al.* (2003) argue that the utilisation of knowledge is not a static inert object to be sent and received and that disseminating new findings or information involves communicating through "certain channels over time among members of a social system" (p. 48). Such a social system can be found within conferences as they provide strategic opportunities for exposure, experience, expertise and embedding strategies, as proposed by Farkas *et al.* (2003), and in doing so they improve knowledge dissemination and utilisation. Their framework demonstrates the gains that can be realised from implementing strategies that are related to each of the opportunities of exposure, experience, expertise and embedding. Those engaged in planning for and delivering conferences can utilise this framework to develop strategies for various stakeholders to advance knowledge and its utilisation.

This chapter's case study has provided examples of a number of useful strategies, however there are many others that could be developed, including:

- ☐ Developing strategies for creating awareness of outcomes of the conference and sharing knowledge and practices;
- ☐ Utilising mentoring to expand experiences with particular practices, sets of skills, or technology;
- ☐ Considering strategies for extending curricula and training programs being held at a conference via hybrid conference methods; and
- ☐ Implementing strategies for embedding new practices or techniques beyond the conference.

The 4 Es are flexible enough to accommodate a focus on a single strategy or multiple strategies.

Further exploration of how the intricate mechanisms which occur at business events characterised by intense knowledge flows (such as conferences) might impact an individual's or an organisation's intellectual and human capital is needed to ensure the potential such events hold is maximised. In a time where we are seeking to address major health and social issues, and being confronted with potentially catastrophic global conditions, events which expertly support and facilitate the generation, acquisition and sharing of knowledge may be key in formulating solutions.

## References

Adler, P. and Kwon, S. (2002) Social capital: Prospects for a new concept, *Academy of Management Review*, **27** (1), 17-40.

Anderson, J. (1983) *The Architecture of Cognition*, Cambridge: Harvard University Press.

Augier, M. and Vendelø, M. (1999) Networks, cognition and management of tacit knowledge, *Journal of Knowledge Management*, **3** (4), 242-261.

Bathelt, H. and Henn, S. (2014) The geographies of knowledge transfers over distance: Toward a typology, *Environment and Planning*, **46**, 1403-1424.

Braun, B. (1992) The economic contribution of conventions: the case of Orlando, Florida, *Journal of Travel Research*, **31**, 32-37.

Caragliu, A. and Nijkamp, P. (2012) The impact of regional absorptive capacity on spatial knowledge spillovers: the Cohen and Levinthal model revisited, *Applied Economics*, **44** (11), 1363-1374.

Centre for Psychiatric Rehabilitation (2021) Boston University College of Health & Rehabilitation Sciences: Sargent College, accessed on 23 March 2021 at: https://cpr.bu.edu/training-and-consultation/4e-framework/

Chua, A. (2002) The influence of social interaction on knowledge creation, *Journal of Intellectual Capital*, **3** (4), 375-392.

Chung-Jen, C. (2004) The effect of knowledge attribute, alliance characteristics, and absorptive capacity on knowledge transfer performance, *R & D Management*, **34** (3), 311–321.

Cohen, W. and Levinthal, D. (1990) Absorptive capacity: A new perspective on learning and innovation, *Administrative Science Quarterly*, 35 (1), 128-152.

Collins, H. (2010) *Tacit and Explicit Knowledge*, Chicago: University of Chicago Press.

Dalkir, K. (2005) *Knowledge Management in Theory and Practice*, Burlington, MA: Elsevier.

Davenport, T. and Prusak, L. (1998) *Working Knowledge: How Organizations Manage What They Know*, Boston: Harvard Business Press.

de Vries, R., van den Hooff, B. and de Ridder, J. (2006) Explaining knowledge sharing: The role of team communication styles, job satisfaction, and performance beliefs, *Communication Research*, **33** (2), 115-135.

Donate, M. and Guadamillas, F. (2010) The effect of organizational culture on knowledge management practices and innovation, *Knowledge and Process Management*, **17** (2), 82-94.

Dwyer, L., Mellor, R., Mistilis, N. and Mules, T. (2000) A framework for assessing 'tangible' and 'intangible' impacts of events and conventions, *Event management*, **6** (3), 175-189.

Easterby-Smith, M., Lyles, M. and Tsang, E (2008) Inter-organizational knowledge transfer: Current themes and future prospects, *Journal of Management Studies*, **45** (4), 677-690.

Edelheim, J.R, Thomas, K., Åberg, K.G. and Phi, G. (2018) What do conferences do? What is academics' intangible return on investment (ROI) from attending an academic tourism conference? *Journal of Dispersion Science and Technology*, **18** (1), 94-107.

Edwards, D., Foley, C. and Hergesell, A. (2016) *Conferences: Catalysts for thriving economies*, report prepared by University of Technology Sydney for Business Events Sydney.

Edwards, D., Foley, C. and Malone, C. (2017) *The Power of Conferences: Stories of Serendipity, Innovation and Driving Social Change*. Sydney: UTS ePress

Edwards, D., Foley, C. and Schlenker, K. (2011) *Beyond tourism benefits: Measuring the social legacies of business events*, Report on behalf of Business Events Sydney, p, 73.

Einebrant, E. (2017) *One day, no one will die from a heart attack*, retrieved from: Destination Jönköping website: http://destinationjonkoping.se/sangeethaiyer/

EuroHeartCare (2017) *Annual congress of the Council on Cardiovascular Nursing and allied professionals: Post-Congress report*, Unpublished internal report.

Farkas, M., Jette, A., Tennstedt, S, Haley, S and Quinn, V. (2003) Knowledge dissemination and utilization in gerontology: An organizing framework, *The Gerontologist*, **43**, 47-53.

Foley, C., Edwards, D. and Schlenker, K. (2014) Business events and friendship: Leveraging the sociable legacies, *Event Management*, **18** (1), 53-64.

Foley, C., Schlenker, K., Edwards, D. and Lewis-Smith, L. (2013) Determining business event legacies beyond the tourism spend: An Australian case study approach. *Event Management*, **17** (3), 311-322.

Foley, C., Schlenker, K., Edwards, D. and Hayllar, B. (2010) *A scoping study of business events: beyond tourism benefits*, report prepared for Business Events Sydney, UTS, Sydney. Available at: http://hdl.handle.net/10453/16848.

Foley, C., Edwards, D and Hergessel A. (2016) *Asian incentive events in New South Wales: expenditure and retail impact*, report prepared for Business Events Sydney, University of Technology Sydney, available at: https://www.besydney.com.au/media/1616/asian_incentive_events_nsw.pdf

Gambina, A., Hagström, J. and Petzold, A. (2018) *Knowledge sharing and organisational development through the 2017 Euroheartcare Conference*. Jönköping University and Destination Jönköping. https://www.themeetingsindustry.org/web/wp-content/uploads/JMIC-Case-Study_Jonkoping_Euroheartcare-2017_18.6.pdf

Grado, S., Strauss, C. and Lord M. (1997) Economic impacts of conferences and conventions, *Journal of Convention & Exhibition Management*, **1** (1), 19-33.

Gupta, A. and Govindarajan, V. (2000) Knowledge flows within multinational corporations, *Strategic Management Journal*, **21** (4), 473-496.

Hanly, P. (2012) Measuring the economic contribution of the international association conference market: An Irish case study, *Tourism Management*, **33** (6), 1574-1582.

Hansen, M. (1999) The search-transfer problem: The role of weak ties in sharing knowledge across organization subunits, *Administrative Science Quarterly*, **44** (1), 82-111.

Hansen, T., Pedersen, D. and Foley, C. (2020) Academic events: An empirically grounded typology and their academic impact, *Event Management*, **24** (4), 481-497.

Hoe, S. L. (2006) Tacit knowledge, Nonaka and Takeuchi SECI model and informal knowledge processes, *International Journal of Organization Theory and Behavior*, **9** (4), 490.

Hughes, C. (1988) Conference tourism – A salesman's dream, *Tourism Management*, **9** (3), 235-238.

Jago, L. K. and Deery, M. (2005) Relationships and factors influencing convention decision-making, *Journal of Convention & Event Tourism*, **7** (1), 23-41.

Jeong, S., Choi, J. and Kim, J. (2014) 'On the drivers of international collaboration: The impact of informal communication, motivation, and research resources', *Science and Public Policy*, **41** (4), 520-531.

Jones, C. and Li, S. (2015) The economic importance of meetings and conferences: A satellite account approach, *Annals of Tourism Research*, **52**, 117-133.

Kim, S., Chon, K. and Chung, K. (2003) Convention industry in South Korea: An economic impact analysis, *Tourism Management*, **24** (5), 533-541.

Kwok, S. H. and Gao, S. (2005) Attitude towards knowledge sharing behavior, *Journal of Computer Information Systems*, **46** (2), 45-51.

Lakoff, G. and Johnson, M. (2008) *Metaphors We Live By*. University of Chicago Press.

Lee, M. and Back, K. (2005) A review of economic value drivers in convention and meeting management research, *International Journal of Contemporary Hospitality Management*, **17** (5), 409-420.

Mair, J. (2013) *Conferences and Conventions: A Research Perspective*, London: Routledge.

Mair, J. and Frew, E. (2018), Academic conferences: A female duo-ethnography, *Current Issues in Tourism*, **21** (18), 2152-2172.

Mair, J. and Jago, L. (2010) The development of a conceptual model of greening in the business events tourism sector, *Journal of Sustainable Tourism*, **18**(1), 77–94

Mair, J. and Thompson, K. (2009), The UK association conference attendance decision-making process, *Tourism Management*, **30** (3), 400–409.

Manipal College of Pharmaceutical Sciences (2017), Sangeetha Iyer interview during the EuroHeart Care-2017, Sweden, Retrieved from: https://manipal.edu/mcops-manipal/mcops-manipal-news/mcops-manipal-news-list/sangeetha-iyer-was-interviewed-during-the-euroheart-care-2017-co.html

Maskell, P., Bathelt, H. and Malmberg, A. (2006) Building global knowledge pipelines: The role of temporary clusters, *European Planning Studies*, **14** (8), 997-1013.

Nonaka, I. (1991) The knowledge-creating company, *Harvard Business Review*, **69**, 96-105.

Nonaka, I. (1994) A dynamic theory of organizational knowledge creation, *Organization Science*, **5** (1), 14-37.

Nonaka, I. and Takeuchi, H. (1995) *The Knowledge-Creating Company: How Japanese Companies Create the Dynamics of Innovation*, Oxford: Oxford University Press.

Nonaka, I., Konno, N. and Toyama, R. (2001) Emergence of 'ba', in Nonaka, I. and Nishiguchi, T. (eds) *Knowledge Emergence: Social, Technical, and Evolutionary Dimensions of Knowledge Creation*, Oxford: Oxford University Press, pp 13-29.

Nonaka, I. and Toyama, R. (2003) The knowledge-creating theory revisited: knowledge creation as a synthesizing process, *Knowledge Management Research & Practice*, 1, 2-10.

Nonaka, I., Toyama, R. and Konno, N. (2000) SECI, Ba and leadership: a unified model of dynamic knowledge creation, *Long range planning*, **33** (1), 5-34.

Oester, S, Cigliano, J., Hind-Ozan E. and Parsons, E (2017) Why conferences matter – an illustration from the International Marine Conservation Congress, *Frontiers in Marine Science*, **4**, 257.

Polanyi, M (1966) *The Tacit Dimension*. London: Routledge & Kegan Paul.

Rutherford, D and Kreck, L (1994) Conventions and tourism: financial add-on or myth? Report of a study in one state, *Journal of Travel & Tourism Marketing*, **3** (1), 49-63.

Severt, D, Wang, Y. Chen, P and Breiter, D (2007) Examining the motivation, perceived performance, and behavioral intentions of convention attendees: evidence from a regional conference. *Tourism Management*, **28** (2), 399-408.

Smith, E. (2001) 'The role of tacit and explicit knowledge in the workplace', *Journal of Knowledge Management*, **5** (4), 311-321

Stacey, R. (2001) *Complex Responsive Processes in Organizations: Learning and Knowledge Creation*, Psychology Press, New York.

Thomas, D. (2006) A general inductive approach for analyzing qualitative evaluation data, *American Journal of Evaluation*, **27** (2), 237-246.

Tsoukas, H. (2005) Do we really understand tacit knowledge?, in *Managing Knowledge: An Essential Reader*, pp. 1-18.

Van den Hooff, B. and De Ridder, J. (2004) Knowledge sharing in context: the influence of organizational commitment, communication, climate and CMC use on knowledge sharing, *Journal of Knowledge Management*, **8** (6), 117-130.

Van Wijk, R, Jansen, J and Lyles, M (2008) Inter-and intra-organizational knowledge transfer: a meta-analytic review and assessment of its antecedents and consequences, *Journal of Management Studies*, **45** (4), 830-853.

Wenger, S. and Snyder, W. (2000) Communities of practice: The organizational frontier, *Harvard Business Review*. Retrieved from: https://hbr.org/2000/01/communities-of-practice-the- organizational-frontier

Zahra, S. and George, G. (2002) Absorptive capacity: A review, reconceptualization, and extension, *Academy of Management Review*, **27** (2), 185-203.

# 6 Legacies through Thoughtful Supply Chains

*Looking around the world I am continuously amazed by the high standards, innovations, and positive impact our industry can make.*

Geoff Donaghy, CEO, ICC Sydney, Group Director of Convention Centres (Asia Pacific) for ASM Global

Prior to the COVID-19 pandemic, more than 13,000 association meetings were being held each year in convention centres worldwide (International Congress and Convention Association [ICCA], 2020). Many of these centres were implementing and documenting new practices based on research conducted into attitudes and outcomes associated with sustainability, including research on:

- The importance of holding meetings at convention centres that incorporate sustainable practices, and the willingness to pay for them (Draper, Dawson & Casey, 2011; Sox *et al.*, 2013; Myung, 2017);
- Green attitudes of convention centre attendees (Lee *et al*, 2013);
- Preference for conferences at venues that make efforts towards sustainability (Buathong & Lai, 2017);
- Trade show delegates' perceptions and willingness to participate in sustainable practices (Jung *et al*, 2016);
- The viability of sourcing local food for meetings and conventions (Lee & Slocum, 2015).

These investigations were a useful starting point for discussion around whether adopting sustainable practices in convention centres would increase demand or delegate satisfaction. Little has been documented in the literature about the potential for convention centres to deliver benefits beyond the venue or event itself (particularly in relation to sustainable food sourcing) and to generate market advantage in so doing.

The International Convention Centre Sydney (ICC Sydney) case study on 'Feeding Your Performance' (Foley *et al.*, 2017; Foley *et al.*, 2018a), which has been adapted in this chapter, demonstrates how a convention centre's significant purchasing power can be utilised to contribute to sustainability by collaborating with, and supporting suppliers in an agri-fresh, short food supply chain. The ICC Sydney Feeding Your Performance program facilitates sustainable behaviour as part of its organisational practices and supports and collaborates with a range of suppliers who are working to improve the agricultural ecosystems in their farming areas. As is the case with sustainable tourism (Liburd, 2018), sustainable business event delivery is dependent upon the viability of the event and the balancing of industry and environmental impacts.

The case study documents the early outcomes of the initiative; a strategy which has enhanced links between ICC Sydney and the agricultural sector and demonstrated the potential of a convention centre to operate as a catalyst for growth, while delivering multiple benefits to delegates. The outcomes of the initiative are discussed in terms of their value for economic, environmental, and social sustainability.

## Case study: ICC Sydney: Feeding Your Performance

| | |
|---|---|
| **Business event(s)** | The case study was compiled utilising research from a number of business events held at ICC Sydney which embraced the Feeding Your Performance initiative. |
| **Location(s)** | Sydney, NSW, Australia |
| **Year(s)** | 2014 to 2017 (ongoing) |
| **Type of event(s)** | This case study relates to an overarching business event delivery strategy applied to numerous events at ICC Sydney. |
| **Event purpose** | The initiative aims to utilise local products and services in the provision of business events to maximise legacy outcomes for local producers and small businesses, the host community and event delegates. |
| **Primary event legacy** | Regional economic support and development for local producers, small businesses and the host community. |
| **Additional legacies (beyond tourism)** | Environmental sustainability, broader community education on healthy eating habits, improved performance of delegates at business events, greater government support, reduced waste production, contribution to charity and diversity promotion. |
| **Summary quote** | "We wanted to have a greater economic impact than just room nights and coffee cups," (ICC Sydney CEO). |

This case study was adapted from: Foley, C., Edwards, D., Harrison, B. and Hergesell, A. (2019)

## Background

ICC Sydney is the largest integrated convention, exhibition and entertainment venue in Australia. It is situated in Sydney, New South Wales (NSW), Australia in the active leisure precinct of Darling Harbour. It is flanked by the Sydney Central Business District and a university precinct. It was opened in December 2016 and employs 1300 staff (300 full time and 1000 casuals).

The United Nations' 70th General Assembly designated 2017 as the International Year of Sustainable Tourism for Development (UNWTO, 2015). The goals include making tourism a catalyst for positive change by promoting socially inclusive and sustainable economic growth, including stimulating employment and encouraging resource efficiency and environmental protection (UNWTO, 2017). ICC Sydney is making a contribution to these goals in three key areas: environmental sustainability, economic sustainability, and social sustainability. This case study demonstrates how these contributions are providing benefits for multiple stakeholders (see Figure 6.1).

**Figure 6.1:** Highlights from ICC Sydney's Feeding Your Performance initiative

| **Environmentally sustainable practices** |
|---|
| Supports local producers and aligns with the NSW government's sustainability strategies; reduces food miles, has a smaller carbon footprint and improves efficiencies; has a focus on recycling, reduced use of chemicals, reduced waste, and effective waste management. |
| **Knowledge sharing** |
| Endorses CSR and stewardship; they are innovative in their aim for restaurant-quality nutritionally balanced meals; includes creative menus and support for new produce. |
| **Economic development** |
| Supports local producers and regional development; provides demand for seasonal fresh produce that counters market volatility inherent in the food industry; further economic growth is anticipated via business expansion, regional growth and tourism. |
| **Social contributions** |
| Encourages awareness of and concerns for farmers and local communities; supports farmers to grow high-quality produce, and new and specialty produce; supports local and diverse communities; nourishes delegates' physical and mental performance via healthy nutritious food. |

## Theoretical and environmental context

Environmental sustainability is concerned with reducing negative human impact and supporting healthy ecosystems necessary for the survival of humans and other organisms (Liburd & Edwards, 2010). ICC Sydney engages in environmentally sustainable behaviour as part of its organisational practices and supports a range of suppliers who are working to improve ecosystems. Key initiatives for environmental sustainability in the agricultural sector in Australia are reducing use of agri-chemicals, managing water scarcity, and maintaining biodiversity (Australian Government, 2009).

*Resource efficiency and waste reduction are important goals for the whole community* (NSW Government, 2014b).

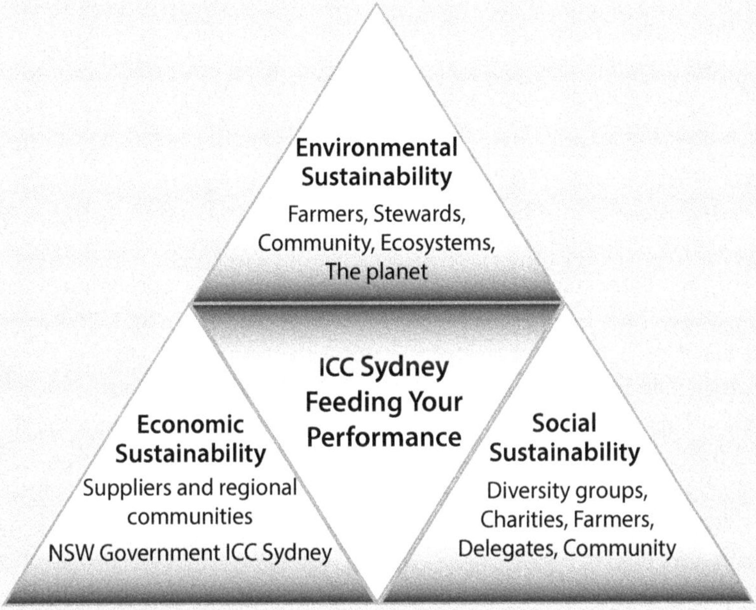


**Figure 6.2:** Key areas of contribution and beneficiaries

Economic sustainability depends on inclusive economic growth, including employment in vulnerable communities (UNWTO, 2017). The agricultural industry is a significant employer in Australia (employing 1.68 million people), particularly in rural and regional areas. The industry contributed $A130.4 billion in food and beverage retail turnover in 2010-11 (Spencer & Kneebone, 2012). Australian food and beverages are highly regarded in terms of quality and food safety, and gastronomic tourism is becoming increasingly important to the tourism industry. Tourism Australia's Restaurant Australia campaign was developed in response to the growing global demand for food and wine as part of the travel experience (Tourism Australia, 2017). However, many New South Wales growers and suppliers in rural and regional areas struggle to compete with low-cost imports, and unemployment is significantly higher than the state average in regional and rural New South Wales (Australian Bureau of Statistics, 2017).

Over the past 20 years, local food has received renewed attention in the western world for a number of reasons. Research has shown local food supply chains can regenerate rural development (Ilbery & Maye, 2005); King *et al.*, 2015), but this requires the hospitality sector to recognise the importance of sourcing locally, to support small business and community livelihoods. ICC Sydney is supporting 65 New South Wales growers and suppliers through sales, publicity and opportunities for expansion, and the program has the potential to influence job growth and regional development. Local food supply chains are seen as the new paradigm to regenerate rural development, not only in New South Wales

6

but globally. An exploratory study of seven European countries, representing 75-85% of the farms in Europe, determined the additional net value generated by local food supply chains on top of conventional agricultural production was positive in all cases (Renting *et al.*, 2003). Similar outcomes have been identified via studies conducted in the United States and the United Kingdom over the past decade (Boyde, 2001; Otto & Varner, 2005; Hughes *et al.*, 2008; Kneafsey *et al.*, 2013; King *et al.*, 2015). Local food strategies have been found to reverse the decline of rural services and the depletion of food and farming physical infrastructure (Kumar *et al.*, 2013).

A growing awareness of the origin of food, food safety and standards, animal welfare issues, and unfair trade has emerged (Renting *et al.*, 2003; Ilbery & Maye, 2005). Moreover, the environmental impacts associated with industrial mass food production and global supply chains have been increasingly criticised (Renting *et al.*, 2003; Ilbery & Maye, 2005; Dunne *et al.*, 2011; Engelseth & Hogset, 2016), with many non-governmental organisations and researchers recommending a reduction in "the fraction of animal-sourced foods in our diets" (Springmann *et al.*, 2016: 4146). There is also general agreement in the literature that a local food supply chain is beneficial for the environment in terms of reduced 'food miles' (Shukla & Jharkharia, 2013) and a smaller carbon footprint (Kneafsey *et al.*, 2013; King *et al.*, 2015). With shorter transportation distances there are generally fewer packaging, processing and refrigeration needs when compared with long-haul industrial food (Kneafsey *et al.*, 2013).

Social sustainability encompasses areas such as equity, diversity and social inclusion, as well as community health, resilience and development (UNWTO, 2017). ICC Sydney is supporting:

- Small food producers to realise their objectives of producing high-quality food;
- Diversity groups through its recruitment and training policies;
- The disadvantaged through its contributions to Sydney charities.

It is also encouraging recognition of farmers and the regional communities in which they operate, while the promotion arising from the Feeding Your Performance program increases community awareness of healthy eating.

According to Chief Executive Officer, Geoff Donaghy, the success of ICC Sydney depends on the success of the events they service. This was the starting point for the development of the working philosophy of Feeding Your Performance, which underpins many aspects of ICC Sydney's service-delivery model, from technological support through to their internal staff management program. However, the Feeding Your Performance philosophy is manifest most notably in the organisation's food and beverage service areas where the aim is for restaurant quality meals to be made from fresh, seasonal and ethically sourced local produce that is nutritionally balanced to energise the physical and mental performance of conference and trade show participants (ICC Sydney Director of Culinary Services).

The Australian agricultural industry plays an important role in delivering safe, nutritious and reliable food products to the Australian people and international trade partners. However, there has been slow growth in the food market since 2008 (Spencer & Kneebone, 2012). Added to this is the issue of ongoing volatility of the operating environment for food producers caused by both climatic and economic conditions. The economic volatility comes from exposure to currency fluctuations that affect export sales, competition from imports, and price fluctuations within or between produce seasons (Spencer & Kneebone, 2012). The cost reduction strategies of the two biggest retailers in the Australian fresh food market, supermarket grocers Coles and Woolworths, have also contributed to the erosion of farmers' margins and many small growers have been forced to exit (IBIS World, 2017). These reports are supported by the suppliers interviewed, who noted that they have been 'squeezed' by the supermarket chains.

To source fresh and healthy local produce, ICC Sydney has identified and partnered with New South Wales farmers, co-ops and providers. Objectives of the direct partnership model are to ensure that ICC Sydney clients receive fresh produce, 'paddock to plate', and to contribute to the sustainability of local and regional small food suppliers in New South Wales. This case study provides an overview of the development of these partnerships, looks at the ways in which the Feeding Your Performance philosophy is permeating the culture of the ICC Sydney operations, and documents some of the outcomes for stakeholders.

## Research methods

This research is based on a qualitative study, the purpose of which was to provide an overview of the outcomes that arise from the Feeding Your Performance program. Interviews were conducted with six New South Wales suppliers at their workplaces in January 2017. Suppliers were identified by ICC Sydney. They comprise:

- A potato farmer from Crookwell
- A cheese maker from Lidsdale
- A lime grower from Kempsey
- A wine maker from Molong
- A butter maker from Tempe
- A seafood supplier from Taren Point.

Personal interviews were also conducted between December 2016 and February 2017 with these seven ICC Sydney executive staff at the ICC Sydney head office:

- Chief Executive Officer (CEO)
- Director of Human Resources
- Director of Finance and Administration
- Director of Communications
- Director of Culinary Services
- Procurement Manager
- Executive Chef.

The final interview was conducted by telephone with ICC Sydney's consultant nutritionist in February 2017.

The interviews, ranging in length from 38 minutes to 182 minutes, were recorded with the permission of participants and subsequently transcribed. The participants were provided with information about the research study and their right to remove themselves from the study at any time.

For purposes of confidentiality, the names of participant suppliers have not been used in this case study. Suppliers are referred to by pseudonyms that represent their area of supply (e.g., potato farmer).

In addition to the interviews, further evidence was collected in the form of relevant media reports, website material and managerial documents. This data and the interview transcripts were analysed using a qualitative data analysis software package (NVIVO). The key themes identified included environmental sustainability, economic development and social impacts.

## Operational aspects of the Feeding Your Performance program

This section provides an overview of what the Feeding Your Performance philosophy entails and how it looks in practice.

Feeding Your Performance is a philosophy, which aims to enhance the performance of:

- ☐ Business event delegates
- ☐ The New South Wales regional economy
- ☐ ICC Sydney.

As part of the program, ICC Sydney partnered directly with 65 New South Wales suppliers, first to ensure that ICC Sydney clients receive fresh produce (paddock to plate) and second to contribute to the sustainability of local and regional small food suppliers in New South Wales.

The performance of attending delegates was enhanced as a result of menu design, which incorporated "seasonal, fresh, nutritious, and healthy food" to provide "the best opportunity for minds to think clearly, engage fully, best ideas, and for bodies to remain light and energised" (Executive Chef, ICC Sydney, cited in Sydney Views 2016). To achieve this, the ICC Sydney engaged a nutritionist who provided advice on menu ingredients that would improve concentration, such as the use of grains that deliver slow-release energy throughout the day.

> *"Feeding Your Performance is all about healthier eating or healthier meal options particularly for our conference guests who might be here for a lunch and then they've got to go back and sit down and still continue to listen and learn ... nutritionally balanced options enable people to still concentrate and take in information"*
> (ICC Sydney Executive Chef).

The performance of the regional economy is supported because the ICC deals directly with the producers. This also means the culinary team are able to learn about the specific ingredients they are working with and the direct access means improved efficiencies and fresher products.

By providing nutritious, energising food to delegates through an exciting menu, and supporting New South Wales industries, the ICC Sydney is also supporting its own performance. Feeding Your Performance is a cornerstone of the ICC Sydney marketing strategy and is successfully being used as a point of difference to promote the convention centre and its offerings (ICC Sydney, 2016).

> *"They [clients] were really excited about what we're doing, and the quality and the standard of the food as well as its presentation is making a powerful impact on their perception of Convention Centre dining and their guests' experience"*
> (Director of Culinary Services).

Having worked in the industry for decades, the CEO believes food and beverage is one of the areas of convention centres that attracts the most criticism and complaints. Thus, it was a strategic decision to address food and beverage offerings from the beginning. To this end, the CEO tasked the new teams to apply an innovative approach to everything they did. The CEO said the philosophy came before the slogan, because he believes people are now seeking health and vitality and good living. However, ICC Sydney soon realised that the slogan and the philosophy had a much broader application as it would be feeding delegates' professional performance as well.

> *"We didn't want Feeding Your Performance to be just another food and beverage slogan. [It is] an underpinning philosophy for everything that we want to do"* (ICC Sydney CEO).

The opening of the new convention centre provided ideal conditions for the Feeding Your Performance philosophy to develop and become embedded in ICC Sydney's operations.

> *"Because our Director of Culinary Services and our executive chef were part of developing the philosophy from the outset, every new culinary recruit has had to adopt this approach"* (ICC Sydney Director of Finance).

She believes it would have been more difficult to take such a philosophy to an established convention centre, stating "I think we've been lucky we did it from day one, so it's just been embedded in ICC Sydney."

The Director of Human Resources confirmed the philosophy permeates the whole organisation. Information about Feeding Your Performance is part of both recruitment and induction processes for all staff members, something that was established by the CEO.

> *"One of the mantras we have here is we're only in business to make our clients successful. That's all. There's no such thing as a successful Convention Centre"*
> (ICC Sydney Director Human Resources).

To progress the philosophy, the executive chef identified and visited a range of farmers and small food supplier producers.

> *"(The aim was) to establish a direct relationship with the people who produced the produce and to share the Feeding Your Performance approach"* (Executive Chef).

6

*"We wanted to have a greater economic impact than just room nights and coffee cups"* (CEO ICC Sydney).

The major stakeholder in the establishment of ICC Sydney was the New South Wales Government. An unexpected outcome of the Feeding Your Performance program is the way it's helping government ministers to deliver on their regional development goals (NSW Government, 2015). The ICC Sydney CEO found the ministers appreciate the support being given to suppliers in regional areas of the state. At the launch of the ICC wine list, Stuart Ayres, (then) New South Wales Minister for Trade, Tourism and Major Events, stated:

*"I know this facility is going to sell over 200,000 bottles of wine every year. That's a huge boost to the local wine industry. A massive endorsement of New South Wales wine and a fantastic way to showcase the high quality and world-class wines we have available"* (News.com.au, 2016).

In summary, Feeding Your Performance – whilst initially designed to strategically address food and beverage considerations – quickly became a philosophy aimed at supporting delegates, other venue patrons, and local and regional suppliers.

# Findings and discussion

This section draws upon responses provided in the interview data and discusses the ways in which Feeding Your Performance engenders sustainability via four broad areas of legacy:

- ☐ Environmental practices
- ☐ Knowledge sharing
- ☐ Economic development
- ☐ Social contribution.

## Environmental practices

The data revealed a number of ways in which the ICC's decision to use New South Wales suppliers ahead of interstate and international alternatives is protecting the environment and effectively reducing the facility's carbon footprint.

### *Sustainability values*

In addition to geographic location, ICC Sydney chose suppliers based on the quality of their produce and their ongoing and proven commitment to sustainable practices.

*"The RFP [request for proposal] pack or the procurement plan [is] very much about making sure the suppliers we adopt or contract share our sustainability thought process. It's almost like interviewing for a job. You either have the chemistry that matches our culture, or you don't. If they're not on-board, they won't reply to our tender"* (ICC Sydney Director of Finance).

Suppliers were vetted for sustainability by ICC Sydney through a comprehensive questionnaire process and, in many cases, through face-to-face interviews and tours of farms and facilities. Suppliers expressed a range of views when it came to sustainability. The potato farmer, for example, had both a pragmatic and an ideological stance, believing in the importance of balancing farming with the ecosystem. Thirty-two per cent of his farmland has been converted to conservation areas including wetlands that support waterfowl habitats. He promotes his practices within the local community to encourage other farmers to adopt similar approaches and hopes to create better outcomes for the community and the whole area.

> *"The best farming practices protect the ecosystem by recycling to reduce waste and use of agrochemicals"* (potato farmer).

The potato farmer also made the point that potato farms in his area (Crookwell, New South Wales) have less reliance on the addition of water, nutrients and insecticides than commercially produced crops grown hydroponically elsewhere in sand using applied nutrients and water. He argues that because his and similar farmers' potatoes are grown in the "actual organic matter in the soil" they take on a richer flavour and have a better texture.

> *"Our soils are very sustainable because they're very old soils and they're rich in organic matter. With our high altitude (1000 metres above sea level) we have cool nights … and altitude also gives you less insect problems. So, we have less reliance on insecticides and fungicides"* (potato farmer).

The wine maker also discussed how important sustainable practices were to his management plans.

> *"I'm the second generation here, we've only owned the property for 38 years, but I've got children who I want to live here. If you are using bad practice, whether it's from in your vineyard, in your winery … it gives you an inferior product. So we recycle everything we can. We compost, we do everything we can, in a manner that just makes sense. Why spray for pesticides if they're not a problem and you've got balance in your vineyard? Why use other chemicals if they're not necessary? It's just common sense… I wouldn't want to drink it if I was lacing it with chemicals, so why would anybody else?"* (wine maker).

As well as growing their own grapes, their sustainability principles extend to their suppliers. The wine maker noted that they only take fruit from local vineyards who adopt similar approaches to their own.

The cheese maker also stated:

> *"we're pretty much sustainable farmers; he [husband] did a diploma in sustainable agriculture after he did his rural degree."*

The Director of Culinary Services gave an overview of her first impression of the lime grower.

> *"He's been in the industry for many years and he's [evolved] his farming practices… when you meet him, his hands are covered in dirt and he's this very passionate grower of chillies and Tahitian limes. The limes … he can grow for 12 months of the year ... he stopped*

6

*growing beans because of the fact that he was having to water them from the top which meant then he was having to use all these sprays to keep the mould off"* (ICC Director Culinary Services).

The ICC was also impressed with the lime grower's focus on hiring locals and buying Australian, stating:

*"Even the glassware he uses for his limes, for his chillies and his chutneys and so forth, it's all Australian made. When I met him I thought, 'wow you're really pretty special'. He's in a really tough industry so listening to him talk about what he's doing with regards to his farming practices and his whole attitude towards Australian produce makes it easy to add him to our supplier list"* (ICC Director Culinary Services).

Over the past eight years the lime grower has also worked to minimise the use of chemicals on the farm.

*"If the grubs want to eat the chillies they can have them ... (laughing) – there is enough to go around. I don't use white oil anymore because with the flowers – the ants eat the aphids and black mould on the leaves can be brushed off with a machine"* (lime grower).

## Fewer 'food miles'

Using local suppliers also means the ICC is minimising the carbon footprint created as a result of their activities, thanks largely to the lower impact of transportation and refrigeration required. One supplier noted that because he is driving at night he doesn't need a refrigerated truck.

*"I can throw the fruit in the back of the truck, leave at 2 o'clock in the morning ... [be] down there by 6-7 o'clock in the morning at his (ICC Executive Chef's) doorstep – drop them off and turn around and be back home again by lunch time"* (lime grower).

ICC Sydney's consultant nutritionist made the point that there are exceptions to the food miles argument, taking rice as an example: "rice is a very water intensive crop, and therefore lots of people have ideas [that] we shouldn't be growing rice in most parts of Australia, because it takes too much water". Dissent for the food miles argument was also voiced by the butter supplier who felt that in a country as large as Australia, dictating food miles can stifle business, particularly in a niche area such as cultured butter.

## Waste management

Effective waste management systems are integral to sustainability for organisations and communities (Shrivastava, 1995). ICC Sydney's contributions to sustainable practice in terms of waste management are numerous. The chefs at ICC Sydney are keen to reduce waste wherever possible, incorporating a number of strategies. First, they have a whole of produce policy.

*"For us we have always been about using as much of the produce [as possible], so for a whole broccoli we try and use the stem, the stalks and the broccoli florets as well so then our waste is minimal"* (ICC Executive Chef).

ICC Sydney is innovative in their approach to food waste, donating unused food to OzHarvest and working with Clean Away to make better use of the recycling system. The waste management company composts the food to make organic fertiliser. Some of this is returned to ICC Sydney for its own rooftop herb gardens.

In summary, ICC Sydney has adopted sustainable practices within its own organisation. It has reduced waste and increased recycling, and is supporting local farmers and small businesses who take sustainability seriously and who adopt stewardship roles in their areas of expertise.

## Knowledge sharing

### *Stewardship*

Stewardship implies service to a community, the responsibility to care for something, including for the benefit of future generations. It is also about accountability and exercising responsibility (Liburd & Edwards, 2010). This section looks at the contribution that ICC Sydney is making in supporting suppliers who take their stewardship roles seriously.

Many of the suppliers understood their role as stewards. The potato farmer explained his interest in sustainability developed as he became older and realised his role as a farmer was less about ownership and profit and more about stewardship, being a responsible custodian of the land he farmed, and improving and protecting the natural ecosystems. He said farming is a "part of his soul" and that money is of little importance to him.

> *"We work because we love what we do and we love being able to produce a good crop … you've got to come to an understanding of the value of what you've got. Not in a monetary sense because how many people have the real luxury of being able to grow food and being able to feed people and take pride in that? We're just custodians of [the land] for our period and we have to care for it in a way that enhances it, ready for the next generation"* (potato farmer).

A similar narrative came from the lime grower who constantly evolves his practices to be more sustainable and shares his experience and knowledge with others, speaking at events and running seminars on farming techniques.

Feeding Your Performance provides ICC Sydney with numerous opportunities to play a stewardship role. The Director of Finance spoke of some of its long-term educational aspects for apprentice chefs.

> *"You don't come here to be a chef. You come here to be sustainable and work on Feeding Your Performance and it's like it's become its own education. You learn [that] we're using the whole carrot. So, the impact to apprentices is huge. They're learning those life lessons forever"* (ICC Sydney Director of Finance).

The finance director also discussed the motivational aspects of Feeding Your Performance with the (non-culinary) ICC Sydney general staff who attend monthly team meetings where pictures of the Executive Chef's visits to regional suppliers are shared.

6

*"So they see (the Executive Chef) there (selecting produce from suppliers' farms) and because he's just one of us, it's that connection too, like they feel like they're actually there picking the mussels ... they know he's famous for mussels ... that Canberra has truffles. People don't necessarily know that these regions that we regularly visit also have wonderful produce. These practices enable staff to say, 'I work for a company that supports this business'"* (ICC Sydney Director of Finance).

The executive chef spoke of the deep knowledge he had gained by liaising with his suppliers, and his belief that this product knowledge enabled him to educate his team and ICC Sydney clients about the produce – to an extent that books could not.

*"All these things that you wouldn't know and I wouldn't have known and now I do know and now my staff know. This knowledge imparted upon us is shared with our staff and clients, providing an opportunity to educate hundreds of thousands of people"* (Executive Chef).

## *Innovation, creativity, agility*

According to the Director of Finance, the Feeding Your Performance initiative is innovative, creative, dynamic and agile. These are important attributes of sustainable programs, organisations and industry sectors over the long term (Edwards *et al.*, 2016):

*"(What the Feeding Your Performance program is doing) is not just for sustainability or CSR responsibility. It's innovative, there is no doubt about it. No-one is doing this right now, particularly from a finance point of view"* (ICC Sydney Director of Finance).

She said they added a component to their Request for Proposal document asking suppliers if they had any new ideas.

*"So, there's a section you can fill in that tells us what we don't know about you that can benefit us. It's amazing the stuff they're putting in there. They're saying things like, 'As well as this service, we also offer this service', and then that opens us up to new ideas... so it's also putting the power back in the supplier's hands to tell us what they're good at, which in turn helps us"* (ICC Sydney Director of Finance).

Staff at ICC Sydney believe that Feeding Your Performance is an innovative approach in terms of serving restaurant quality food in a convention centre, which is nutritionally balanced and energising.

*"I don't know how many conventions I've sat in where you get the same stuff for morning tea, lunch, afternoon tea, and by about three-thirty, you just want to have a sleep under the table, because you've eaten far too much carb ... how that will flip from a client point of view, I think that's exciting to watch"* (ICC Sydney Director of Finance).

The executive chef also feels that the concept of a convention centre providing restaurant quality food breaks new ground and is something "no one's ever dreamed of before." He said the close relationship established with suppliers even leads to new products being developed. For example, they are working with the potato farmer to develop a variety that will crisp without the use of oil and turn orange. The executive chef sees this as a collaborative journey in delivering wholesome, nutritional food. By committing to purchasing these potatoes (once developed), ICC Sydney is enabling the potato farmer to invest in developing new produce lines with less vulnerability to risk.

In summary, Feeding Your Performance is an innovative initiative that is acting as a catalyst for further innovation through its approach to relationships with suppliers, creativity in menus and corporate social responsibility. These actions further engender sustainability practices that align with the New South Wales Government's sustainability strategies (2014, 2015).

## Economic development

### *Supporting local producers*

The emphasis on ensuring products are supplied by local (New South Wales) businesses also creates legacy benefits for the local economy. This was seen as an important factor in all interview responses.

> *"I love that Tony [Executive Chef] has embraced using local producers. He's really … helping them improve their businesses, as well as them providing really great produce for him, so I think that's a real plus, and essentially that's what real food's all about … supporting our agricultural producers, as well as those who are [contributing] through making foods like cheeses, and pickles, and relishes, and popular butchers, great fishmongers. That's all part and parcel of healthy eating … not just for us, but also for the country and for the planet"* (consultant nutritionist).

The volatility of the operating environment for Australian food producers is not expected to decrease in the near future and, in this context, ICC Sydney's strategy to provide a level of stability for their food suppliers is a significant contribution. In addition to supporting individual suppliers, ICC Sydney is contributing to a growing set of farmer and consumer-led initiatives which include: farmers' markets; marketing cooperatives; community supported agriculture; and direct and online sales hubs (Estrada-Flores & Larsen, 2010). The seafood supplier has noted the growth of this movement.

> *"It has opened up people's eyes to what food is … and maybe [future generations] will start demanding Australian produce … the emergence of the little operators is coming back again. Retail for small operators has boomed because people, as time poor as they are, will go to a butcher … they'll go down to the fish markets … it's better than where it was five years ago"* (seafood supplier).

Patronage by quality-conscious customers such as ICC Sydney provides an opportunity for small food production enterprises to increase stability in a volatile market and to grow their customer base.

While the partnerships between suppliers and ICC Sydney are new, suppliers anticipate several benefits. Expected job growth may be small, but suppliers feel the impact on regional towns will be meaningful. Other benefits include business expansion, regional promotion and increased tourism opportunities. As the potato farmer stated, "If it employs one more person in this little town well that's a bonus".

Image provided by ICC Sydney

## *Supporting local small business*

The nutritionist also noted Feeding Your Performance is helping small, local businesses.

> *"I know a couple of bakers (we're) working with. [They] were very small. He's helping them build their business … he's got to work with more than one, because a small producer can't produce the kinds of quantities the Convention Centre is going to need, but he's allowing them to grow their business, and of course, that's keeping the money in New South Wales, allowing New South Wales producers to profit"* (consultant nutritionist).

A number of the suppliers are using their relationship with ICC Sydney to plan for future growth. As noted previously, the potato farmer is extending his business and others in his region by growing a specialty 'ware' potato just for ICC Sydney. The cheese maker is using the increase in sales to consolidate and improve cash flow.

Such activities are possible because ICC Sydney has allowed suppliers to set their own payment terms, with some choosing a quick turnaround of 15 days or the end of the month. However, for the butter maker the indirect benefits are of greater significance.

> *"The video that they did – the exposure … I guess where we're at with our company at the moment is exactly that. We need growth but we need growth in a different way … the public relations and marketing that ICC Sydney undertakes has the biggest impact on our business today"* (butter maker).

By having their produce on the ICC Sydney tables means their butter is on the world stage. They are using the relationship with the convention centre to get to "that next level", particularly through public relations and world-wide exposure via international visitors to the convention centre. ICC Sydney supports this approach and promotes suppliers through marketing and during events. In this way the suppliers become part of ICC Sydney's "messaging, and storytelling [which they feel] deeply personalises the whole process" (ICC Sydney CEO).

It is a similar story for the sustainable seafood supplier, who is now able to boast he "doesn't just cater for small restaurants ... or two chef hat restaurants - we also supply businesses like the ICC Sydney". From his perspective, if ICC Sydney is buying from them, then others will be confident he can deliver.

> *"I'm proud in telling them we're doing ICC Sydney. Every time I tell someone, I tell them these guys aren't your usual volume event space"* (seafood supplier).

The artisan cheese makers (husband and wife team) have won numerous awards for their goat cheese products and have some high-end clients. However, they have been struggling to sell enough volume. They stated they are very pleased to partner with ICC Sydney because it means they can focus on making cheese products (their forte) and spend less time looking for new customers.

Image provided by ICC Sydney

> *"We've got good clients and good customers – we need more, because we need [to make a living], you can't keep nagging the [customers you have] ... It's hard work marketing"* (cheese maker).

Partnering with ICC Sydney allows suppliers to focus on their core competencies of delivering high-quality produce. All suppliers highlighted the opportunity for regional development as a key benefit they hoped would eventuate.

The wine maker noted that, while 'paddock to plate' has become popular in terms of sustainable practice and promoting high-quality restaurant food, the same cannot be said for beverages.

> *"There are some fantastic restaurants in Sydney that will sit you down and they'll tell you about the food that 'comes out of the garden here and we've got our own farm at Windsor and we pick everything there,' and then ... 'Would you like this* French *shiraz to have with that?"* (wine maker).

He is hopeful the industry rhetoric about paddock to plate will eventually translate to beverages and that his relationship with ICC Sydney will expedite that shift. To this end, ICC Sydney is contributing to a newly emerging movement towards a well-deserved appreciation of New South Wales wines. The Director of Culinary Services explained they "were very aware from the outset the impact the ICC Sydney wine list could have for that

6

winemaker" from a volume and sales perspective as well as for regional brand awareness. Because of this. the sommelier was tasked with building an 80 per cent New South Wales wine list, which he is very passionate about.

> *"If you look around Sydney's dining establishments, you rarely see New South Wales wines. There's 14 regions now in New South Wales going from cooler climates right through to the Hunter Valley ... normally in big venues and in hotels your house wine or your house package is done on price ... but we want to offer our guests quality wine, [regional] wine labels to be listed"* (ICC Sydney Director of Culinary Services).

Image provided by ICC Sydney

The wine list was built through a blind tasting of just over 1300 New South Wales wines, ensuring every wine earned its place on the wine list for quality and not price. ICC Sydney believe this is an industry first. The strategy is paying off by way of positive feedback and promotion of New South Wales and smaller wineries. Since the opening of ICC Sydney, the Australian wine industry has held two events there, focusing on New South Wales wine labels and aimed at sommeliers, wine buyers, critics and the media. One event largely used the ICC Sydney wine list of labels for its master class tasting.

In summary, ICC Sydney's strategies are contributing to jobs in urban, rural and regional New South Wales, regional development and branding, and providing hope for stability and growth to small food producing businesses in this volatile environment.

## Social contributions

Local food supply chains can strengthen relationships between members in food chain communities. This is particularly true for farmers (Mundler & Laughrea, 2016). Local food promotes social and professional recognition for farmers and may even allow for a form of social and professional reintegration for vulnerable or marginalised farmers (Mundler & Laughrea, 2016).

Further, a focus on local food arouses social concern for farmers. Studies have shown that people buy local food, to a large extent, because they feel sympathy for the farmers (Corsten & Felde, 2005; Arsil *et al.*, 2014; Mundler & Laughrea, 2016). Empathy and respect for farmers was evident in all the interviews conducted for this case study.

*"I think it's a good story that we're supporting local communities like that, and they're doing it tough out there for whatever reason, and if it creates wealth and jobs and things like that for the families it's great"* (ICC Sydney Procurement Manager).

## *Supporting farmers to grow high-quality produce*

ICC Sydney's Executive Chef is sourcing produce with high nutritional value and flavour. He explained:

*"Remember the way tomatoes used to taste before they were modified for long shelf life and to look good in supermarkets? I've found someone who grows heritage tomatoes. I don't even like to call them heritage, they are just real tomatoes, full of flavour"* (ICC Sydney Executive Chef).

The potato farmer provided a detailed explanation of a similar issue in the potato industry where intensive farming is now a problem when producing high-quality produce.

*"If you compare the fruit and vegetable today with what we had 30, 40 years ago they don't have the same level of nutrient in them and that's exactly what I was saying about growing in the sand … sand is inert, the vegetable doesn't take the zinc and all the things that are in soil ... we're going back to looking at more sustainable land and smaller areas that will produce a much higher quality nutritional food"* (potato farmer).

The potato farmer also suggested the ICC Sydney consider niche potatoes, such as the wild potatoes that come from the Andes, as they are better nutritionally and "ooze flavour".

*"If you want to buy commodity potatoes you just go to the supermarket and buy whatever's cheap. But if you want to have a meal experience … where you want to know the story of your food, you want to enjoy what you eat and it's a complete experience, then that takes a whole different strategy"* (potato farmer).



## *Supporting the local community*

ICC Sydney supports the local community through recruitment of diverse groups, including people with learning disabilities, the LGBTI community, youth technical colleges and hospitality schools, working mothers and mature workers. They have achieved this by partnering with diversity organisations that focus on specific demographics, in order to get their messages out (ICC Sydney Director of Human Resources).

They also work with OzHarvest, which collects leftover goods and distributes them to people in need and the homeless. OzHarvest also works with young people, providing opportunities to find employment by connecting them with organisations like ICC Sydney. ICC Sydney has further plans to support charitable organisations through an employee charity work scheme.

*"ICC Sydney will partner with charity organisations on a number of fronts including where we offer a day of charity work to every employee. Whatever we decide to do we can link in with those organisations to say right do you need help with x, y and z? We can give out this many staff to help you"* (ICC Sydney Director of Human Resources).

Many such local food schemes have been shown to increase community knowledge and behavioural change around healthy eating (Kneafsey *et al.*, 2013; Mundler & Laughrea, 2016). ICC Sydney's Feeding Your Performance program is contributing to a new level of nutritional awareness in three ways. First, they are supporting growers who are working towards producing food with exceptional nutritional value. Second, they are providing delegates with nutritionally superior food and information about local producers. Finally, they are donating any excess of this high-quality food to an organisation that provides meals for people in Sydney who are experiencing disadvantage.

### *Nourishing performance*

The benefits associated with providing nourishing food to delegates in order to improve their performance and experience of business events at ICC Sydney has been an important motivator. ICC Sydney also supports the New South Wales Government's Healthy Eating and Active Living Strategy (2014a). Engaging a consultant nutritionist has further assisted it to realise this objective.

> *"We discussed various philosophies that should be on the menu; to control energy levels for people [who are] there all day, for helping brain power throughout the day ... for concentration and to stop people having that struggle to keep your eyes open"*
> (ICC Sydney's consultant nutritionist).

Increasing performance is based on minimising carbohydrate portions and lowering the glycemic index (GI). This enables "energy to trickle into the bloodstream through the afternoon, helping to feed your brain, helping you to keep your brain power and concentration, and not allow you to have quite such a big slump in the afternoon" (ICC Sydney's consultant nutritionist). As such, there is a stronger focus on the use of whole grains and legumes, quinoa, and a variety of newer grains. The role of the nutritionist has been to review every single recipe being used by the culinary team and guide them in menu preparation.

> *"They are on the right track with the sorts of things they are using in the recipes, so the meals are light, high in protein and use low GI carbohydrates, wholegrains, minimally processed grains and less sugar. The executive chef and I are exactly on the same page with regards to using real food and using top quality produce, and ... well it's not rocket science ... (you just need to) get a really nice balance"*
> (ICC Sydney's consultant nutritionist).

Research has found a strong link between diet and sustainability. In particular, dietary changes that involve increased consumption of plant-based foods and reduced consumption of animal foods lead to a reduction in greenhouse gas emissions (Scarborough et al. 2014; Stoll-Kleemann & O'Riordan, 2015; Springmann *et al.*, 2016). The consultant nutritionist believes there is not enough information available to consumers to understand this issue and feels the Feeding Your Performance program is helping to convey the messages about eating more plant foods, oysters and mussels, seafood that is much more sustainable. She believes eating more plants "will have a huge influence on the health of our planet."

## Conclusions

The impact of ICC Sydney's Feeding Your Performance approach to venue management is an important story for our industry and a good example of the impact a convention centre can have upon the legacy benefits for its local and regional community. The economic impact of business events is well documented via hotel room nights, dining and shopping, visiting attractions, local transport and the like. Now the industry is maturing to the point where it is looking more deeply at how it measures the benefits beyond tourism.

> *"By measuring the impact of Feeding Your Performance we're tracking delegate experiences, measuring our corporate social responsibility performance and embarking upon an entirely new area. [Things like] the economic impact of our venue on its supply chain and the communities they operate in - in terms of healthy businesses and jobs growth for example"* (ICC Sydney Director of Communications).

This case study has reported on the benefits and impacts of the sustainable and inclusive practices adopted by ICC Sydney, as part of its Feeding Your Performance program. ICC Sydney is making contributions in three key areas (sustainability, economic development, and social impacts) and providing benefits for multiple stakeholders.

### *Environmental sustainability*

ICC Sydney engages in environmentally sustainable behaviour as part of its organisational practices and supports a range of suppliers who are working hard to improve ecosystems. ICC Sydney suppliers benefit directly from this support. Their respective industries benefit from the stewardship shown by many of these suppliers. The broader community benefits from reduced waste and fewer chemicals in their food, and the planet benefits from improved ecosystems.

### *Economic development*

ICC Sydney is supporting 65 New South Wales suppliers to grow their businesses through sales, publicity and opportunities for expansion. The program has the potential to influence job growth and regional development, and businesses are being encouraged to develop new products.

### *Social contributions*

ICC Sydney is supporting:

- ☐ Small food producers to produce high-quality, affordable food in sustainable ways;
- ☐ Diversity groups through its recruitment and training policies;
- ☐ Healthy communities through education and stewardship on nutritious eating habits;
- ☐ The disadvantaged through its contributions to Sydney charities.

It is also providing recognition, promotion and reward for farmers and the communities in which they operate.

### Alignment with government strategy

Finally, ICC Sydney's initiatives align with various New South Wales government policies (Economic Development Strategy for Regional NSW, the Healthy Eating and Active Living Strategy and sustainability strategies) in supporting regional suppliers and communities and promoting healthy eating.

# Theoretical discussion

## Corporate Social Responsibility (CSR)

In the latter years of the twentieth century consumer concerns were raised around the ways in which some corporations conducted themselves and the negative impacts their practices might have upon wider society. The concept of CSR emerged as a distinct field in the early 1990s (Blowfield & Frynas, 2005, p. 194) and several different definitions have been put forward to explain the term (for example, Blowfield & Frynas, 2005; Tikkanen, 2012; Gold & Heikkurinen, 2013). The core idea of CSR is that stakeholders, including customers, suppliers, local communities, non-government organsiations, employees, government, investors and other supply chain members (Maloni & Brown, 2006), may hold a corporation socially and ethically responsible for their actions. CSR includes a responsibility companies have for their impact on society and the natural environment (Blowfield & Frynas, 2005) and for ensuring sustainable development (Ben Youssef *et al.*, 2017). These definitions suggest accountability, transparency, competitiveness and responsibility are important parts of CSR (Gold & Heikkurinen, 2013).

Food companies and the way they source and transport their produce are an important focus of CSR attention. A number of motivations and benefits which might prompt a company involved in food supply and distribution to adopt ethical and socially responsible practices have been identified, including that they can:

- Enhance a company's competitive advantage (Porter & Kramer, 2006)
- Improve staff morale, better manage risk, increase profits, market share and reputation, and strengthen value chains through responsible sourcing (Arato et al., 2016)
- Provide access to global markets (Porter & Kramer, 2006; Manning, 2013)
- Improve social image (Smith, 2003)
- Reduce costs as a result of sustainable waste disposal actions such as reusing, recycling and waste management (Wiese & Toporowski, 2013).

Organisations can face a range of challenges when implementing CSR practices (Harrison et al., 2019). These challenges include finding ways to demonstrate tangible economic benefit to top management from CSR activities (Székely & Knirsch, 2005) and the need to acquire additional knowledge and resources to implement CSR activities which are otherwise outside a company's area of expertise (Freeman & Liedtka, 1991; Hemphill, 1997). Furthermore, many consumers' enduring preference for buying products based on price over consideration of CSR practices makes it challenging for many organisations to decide which CSR activities should be chosen and how they might influence customers' buying decisions (Hemphill, 1997; Smith, 2003).

Despite continued demand for competitive pricing of products, customers are increasingly indicating stronger intent to buy from organisations with proactive CSR initiatives (Kim, 2017). They are also increasingly expressing concern about the way products are procured, produced and distributed (Costanigro *et al.*, 2016). This encourages responsible sourcing, an important aspect of sustainable supply chain management (Zorzini *et al.*, 2018), which can further contribute to social, environmental and economic benefits for communities.

Finally, development of surrounding communities can be realised through local sourcing and support for economic development (Pullman & Dillard, 2010; Foley *et al.*, 2018b).

## Short food supply chains

Kneafsey *et al.* (2013) defined an agri-fresh short food supply chain as one where "the foods involved are identified by and traceable to a farmer" and where the number of intermediaries between farmer and consumer is minimal or ideally nil. A short food supply chain is driven by a unique set of characteristics. These characteristics include the perishable nature of produce and continuous degradation of quality; long lead times for producing and bringing perishables to market; seasonality in production and consumption of perishables; and reduced numbers of intermediaries between the farmer (food producer) and the consumer (Harrison *et al.*, 2019). The number of (or absence of) intermediaries is often indicated as a discriminating factor in defining short food supply chains (Jarzębowski et al., 2020, p. 3). Too many actors in the supply chain, however, can make a food supply chain highly complex and less transparent, and consumes more resources in the process of delivering produce to customers (Harrison *et al.*, 2019). The concept of the local or short food supply chain has evolved to address these specific issues and to make the agri-fresh food supply chain more sustainable, economically, environmentally and socially (Enjolras & Aubert, 2018).

According to Marsden *et al.* (2000), short food supply chains re-socialise or re-spatialise food, which allows "the consumer to make value-judgements about the relative desirability of foods on the basis of their own knowledge, experience, or perceived imagery" (p. 425). Marsden *et al.* used the terms re-socialise and re-spatialise to indicate the locality of the food produced, to draw upon and enhance an image of the farm and to redefine the producer-consumer relationship based on the food's origin. They argue that this relationship plays an important role in short food supply chains as it is the means by which value and meaning are constructed rather than what is created by the produce alone (Marsden *et al.*, 2000). They propose three distinct types of short food supply chains:

1. Face-to-face: in which customers directly interact with farmers and buy products;
2. Spatially proximate: where customers buy products from local outlets where all locally produced products are aggregated and sold;
3. Spatially extended: where customers located away from the region buy products grown in a particular region based on the labels and key information available about the product.

The central idea is that short food supply chains do away with the need for long, complex and rationally organised industrial chains (Marsden *et al.*, 2000), thereby providing locally produced, fresh food that is an attractive option for customers (Harrison *et al.*, 2019). The attraction of the short food supply chain can be increased by creating awareness about the locality of the product and by explaining the benefits of consuming that product over more large-scale, imported or commercially produced options. These benefits include freshness, transparency in the process, provenance, reduced environmental impact and cost due to fewer food miles, increased job opportunities for the local population, and support for local organic and small-scale growers (Feagan *et al.*, 2004; Kottila & Rönni, 2008; Brunori *et al.*, 2016; Harrison *et al.*, 2019). In addition, short food supply chains have been found to exert positive economic influences on their local region, mainly in terms of economic growth and employment (Ilbery & Maye, 2005). They are seen as a new paradigm to regenerate rural development globally.

There are, however, several challenges to short food supply chains (Harrison *et al.*, 2019). Produce tends to have a higher price due to small-scale production, policy and regulatory constraints, and the high cost of registration for certification (Oglethorpe & Heron, 2013). These high costs thus require additional value in terms of quality and freshness (Lehtinen, 2012), which means local produce can only be supplied in season, restricting customers from relying solely on local supply (Feagan *et al.*, 2004). However, local suppliers find it difficult to compete with large-scale suppliers and to penetrate retail chains and institutions like airports, large hotels and cruise terminals, because they cannot guarantee supply and a cost per unit that would reduce the total cost of purchasing (Foley *et al.*, 2018b; Harrison *et al.*, 2019).

Organisations and corporations which genuinely embrace CSR philosophies, however, are able to work around such challenges, and supplier development is a viable approach for buying firms to improve the performance of the supply chain (Krause *et al.*, 2000, 2007; Foley *et al.*, 2018b); Harrison *et al.*, 2019). Buying organisations may adopt strategies to improve the capabilities of their suppliers and, in so doing, address community development and improve suppliers' economic performance. Supplier development is viewed as an important corporate social responsibility strategy in that it aims to develop supplier capabilities and advance social performance via improved economic performance (Yawar & Seuring, 2018; Harrison *et al.,* 2019). The case study presented in this chapter offers a useful example of how such goals may be achieved, for example, ICC Sydney actively supported the potato farmer in developing new varieties of crops by guaranteeing sales once achieved.

Economic development depends on inclusive economic growth, including creating and supporting employment in vulnerable communities (UNWTO, 2017). In 2018-2019, the Australian agricultural industry contributed 2.2% of value added (GDP) and accounted for 2.6% of national employment (Jackson *et al.,* 2020). Australian food and beverage is highly regarded in terms of quality and food safety, and gastronomic tourism is becoming increasingly important to the tourism industry (Tourism Australia, 2017). However, many New South Wales growers and suppliers in rural and regional areas struggle to compete with low-cost imports, and unemployment is significantly higher than the state average in regional and rural New South Wales (Australian Bureau of Statistics, 2017). Initiatives like those highlighted in the case study operate against the aforementioned social problems and contribute to thriving local economies.

## Sustainable procurement through CSR



Support for agriculture and local farming entities through sustainable procurement programs is a cornerstone of ethical social behaviour for corporations involved in the provision of food. According to Leire and Mont, sustainability is concerned with "the utilization of the purchasing power of public and private organisations to purchase products, works and services that have positive social impact on the society" (2009, p. 29). The principles of sustainable food procurement include:

- ☐ The sourcing of raw local materials required for production (Lehtinen, 2012);
- ☐ Designing menus with resource consumption details of the product to create awareness among consumers about the food they eat (Bryan & Middlecamp, 2017);
- ☐ Taking steps to reduce purchase of non-sustainable foods, collaborating with others for centralised purchasing systems, ensuring resource-efficient processes are followed throughout the production stages, and utilising efficient transportation to reduce food miles and the corresponding environmental impacts (Cerutti *et al.,* 2016)
- ☐ Sourcing animal foods that are produced by following national regulations, food prepared with minimum amounts of additives, and adoption of corporate codes for addressing issues embraced by international labour organisations (Rimmington *et al.,* 2006; Tikkanen, 2014).

Sustainable procurement through CSR and the concept of the short food supply chain are evolving globally. However, many farmers still face significant challenges when attempting to penetrate large tourism and hospitality organisations such as airports, hotels and cruise terminals (Harrison, *et al.*, 2019). This is because these institutions procure their food products from large suppliers to ensure continuous supply and reduce purchasing costs. Whilst this desire to achieve greater certainty and convenience is understandable, it can also limit an organisation's capacity to benefit from the many advantages and opportunities short food supply chains can offer.

To source fresh and healthy local produce, ICC Sydney identified and partnered with New South Wales farmers, co-operatives and providores (specialty suppliers) who shared a similar focus for sustainability. The case study documents their stories and explains how the collaborative partnerships between these suppliers and ICC Sydney are contributing to the key areas of environmental sustainability, knowledge sharing, economic development, and social development.

## Lessons learned

The case study presented in this chapter demonstrates how a procurement program implemented at a large international convention centre located in Sydney impacted small businesses and suppliers throughout the regional community. The convention centre developed a program, in which a unique short food supply chain of around 60 local and regional farmers work in partnership with the convention centre to provide fresh local seasonal produce.

ICC Sydney took a proactive approach with local producers, enabling them to enter the supply chain on terms that were best for their business and adjusting ICC Sydney's in-house logistics to support the delivery needs of suppliers.

Hjalager (2018) presents a model of suppliers' roles and capacities in sustainable tourism (FIgure 6.2). In many respects, the case study demonstrates the benefits highlighted in Hjalager's model, which arise when a firm takes a proactive, co-developing and/or co-branding role with their suppliers, preferably with direct interface and personal contact.

This case study presents strong evidence of how the direct interface plays out in practice. ICC Sydney is formally collaborating with suppliers to achieve their sustainability agenda. Advancing their knowledge of each other because of an emphasis on people-to-people interactions, the gains are fruitful for both parties. The ICC Sydney Feeding Your Performance program is a clear example of an events/tourism entity, with a distinct sustainability agenda, actively in search of collaborative suppliers and enabling those suppliers to come up with solutions (Hjalager, 2018). It demonstrates that in tourism, collaboration and partnerships *can* form a basis for resilient action (Foley *et al.*, 2018b).

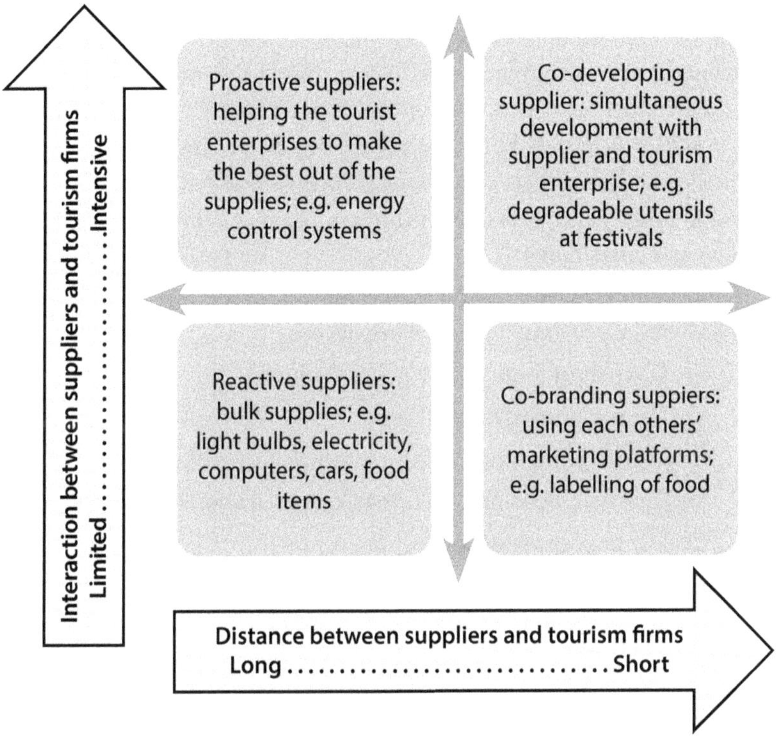


**Figure 6.2:** Model of suppliers' roles in sustainable tourism development. After Hjalager (2018)



Another important message to be drawn from the case study is that, whilst the operators of ICC Sydney are very much aligned with, and committed to, the ethical, responsible and community-minded practices adopted in their Feeding Your Performance program, they are also responsible for the ongoing financial viability and profitability of the convention centre. But rather than see these two concepts as being mutually exclusive, they have adopted a strategy which serves both objectives. They have designed a program which not only aligns with responsible and ethical sourcing of supplies to support their local economy, but which also improves ICC Sydney's competitive advantage by enabling them to differentiate their brand in a highly competitive global marketplace. Importantly, it also has enabled them to extend their circle of influence with important stakeholders because their preference for New South Wales suppliers supports the goals of New South Wales government ministers looking to bolster regional development. The more ICC Sydney is seen to be contributing to the prosperity of regional producers and economies, the more value they are seen to be adding to the regional economy; hence the more community and political support they are likely to attract.

In addition to this, they maintain that by providing outstanding culinary experiences as a result of their direct food and beverage procurement strategy, they improve their clients' performance, outcomes and satisfaction. The produce they source supports the culinary aspect of the program where ingredients are combined to enhance the physical and mental performance of delegates attending events at the convention centre. This further differentiates them from competitors and provides an added value proposition to be used in marketing.

Whilst many of these motives are far from altruistic, they do demonstrate there are legacies that can be created via business event delivery which can have a broad range of impacts likely to bolster an event's (or a venue's) likely successfulness and sustainability.

In summary, ICC Sydney's Feeding Your Performance program and CSR initiatives are having a positive impact on:

- ☐ Their short food supply chain suppliers
- ☐ The regional economy
- ☐ Local communities
- ☐ Disadvantaged and marginalised groups in the local community
- ☐ The local environment
- ☐ Delegate outcomes and satisfaction
- ☐ ICC Sydney's long-term sustainability
- ☐ ICC Sydney's reputation in the community and within government
- ☐ Staff morale and satisfaction
- ☐ Their profitability
- ☐ Other hospitality providers in neighbouring communities which are taking their lead.

(Foley *et al.*, 2017; Foley *et al.*, 2018a; Foley *et al.*, 2018b).

It should be noted, however, that these types of initiatives are dynamic in nature, have their own inherent risks and may each have very different characteristics depending on the people involved and local circumstances. Whilst this case study highlights important benefits and outcomes from a program strategically designed to support a range of stakeholders in New South Wales, it is a relatively new initiative, the long-term viability of which is not yet clear. In terms of the case study, there is the issue of what might happen if there were changes in key roles at ICC Sydney. Would new incumbents embrace the sustainability agenda as much as the current ones do? If not, what would happen to the suppliers who had reconfigured their practices to meet the needs of this major client? Maintaining collaborative relationships and knowledge continuity when key people retire or move away is significant to ongoing collaborative programs (Carlsen & Edwards, 2013).With the program still so new, there is uncertainty around whether collaborations will deepen over time, whether the expectations of suppliers will continue to be met, and how approaches by prospective new suppliers will impact relationships and be managed.

Such issues of risk warrant consideration and it is important that strategies are put in place to minimise them. For example, ICC Sydney could undertake to ensure they provide adequate advance warning and clear communication about any plans to terminate relationships. This would give producers time to identify other markets and to adjust their business practices.

Finally, whilst feedback and commentary indicate the Feeding Your Performance program does in fact improve delegate performance, thus far the claim has not been documented via formal research studies. Such studies would add weight to the program's value proposition and relevant marketing potential.

These issues aside, the approach taken by the convention centre in their procurement program provides a blueprint for short food supply chains that could be adopted by other large venues and institutions such as hospitals. State and national governments could foster similar strategies in government-run venues to support the producers in their own regions, in turn, contributing to government objectives for economic support of rural and regional areas. Commercial venues (including convention centres) in urban centres could support their regions by purchasing and showcasing high-quality local produce, thereby enhancing their own credentials for superior food and beverage offerings, as well as demonstrating a commitment to corporate social responsibility.

For this to occur, further research would be required around the barriers and challenges to a range of businesses in the hospitality sector adopting a short food supply chain approach. We also suggest that empirical research conducted in partnership with industry and government that is focused on short food supply chain policies could contribute to building stronger urban-rural linkages between consumers and their food.



## References

Arato, M., Speelman, S. and Van Huylenbroeck, G. (2016) Corporate social responsibility applied for rural development: An empirical analysis of firms from the American Continent. *Sustainability*, **8** (1), 1-17. https://doi.org/10.3390/su8010102.

Arsil, P., Li, E., Bruwer, J. and Lyons, G. (2014) Exploring consumer motivations towards buying local fresh food products, *British Food Journal*, **116** (10), 1533-1549.

Australian Bureau of Statistics (2017) New South Wales (S/T), viewed 20 March 2017. https://www.abs.gov.au/statistics.

Australian Government (2009) Feeding the future: sustainable agriculture, viewed 30 April 2017. http://www.chiefscientist.gov.au/2009/10/feeding-the-future---sustainable-agriculture/

Ben Youssef, K., Leicht, T., Pellicelli, M. and Kitchen, P. (2017) The importance of corporate social responsibility (CSR) for branding and business success in small and medium-sized enterprises (SME) in a business-to-distributor (B2D) context, *Journal of Strategic Marketing*, 1-17.

Blowfield, M. and Frynas, J. (2005) Setting new agendas: critical perspectives on corporate social responsibility in the developing world, *Journal of the Royal Institute of International Affairs 1944*. https://doi.org/10.2307/3569630

Boyde, T. (2001) *Cusgarne Organics local money flows*. New Economics Foundation and The Countryside Agency.

Brunori, G., Galli, F., Barjolle, D., van Broekhuizen, R., Colombo, L., Giampietro, M., et al. (2016) Are local food chains more sustainable than global food chains? Considerations for assessment, *Sustainability*, **8**, 1-27. https://doi.org/10.3390/su8050449

Bryan, T. and Middlecamp, C. (2017) Learning through eating: bringing campus dining operations into an Environmental Science Course, *Sustainability*, **10** (1), 30-38. https://doi.org/10.1089/sus.2017.29081.twb

Buathong, K. and Lai, P. (2017) Perceived attributes of event sustainability in the MICE industry in Thailand: a viewpoint from governmental, academic, venue and practitioner, *Sustainability*, **9** (7), 1151.

Carlsen, J. and Edwards D. (2013) Diablo Trust: planning sustainable land use, in J. Liburd, J. Carlsen & D. Edwards (eds), *Networks for Innovation for Sustainable Tourism: Case Studies and Cross-Case Analysis*, Melbourne: Tilde University Press, pp. 41-48.

Cerutti, A., Contu, S., Ardente, F., Donno, D. and Beccaro, G. (2016) Carbon footprint in green public procurement: policy evaluation from a case study in the food sector, *Food Policy* **58**, 82-93. https://doi.org/10.1016/j.foodpol.2015.12.001

Corsten, D. and Felde, J. (2005) Exploring the performance effects of key-supplier collaboration: an empirical investigation into Swiss buyer-supplier relationships, *International Journal of Physical Distribution & Logistics Management*, **35** (6), 445-461.

Costanigro, M., Deselnicu, O. and McFadden, D. (2016) Product differentiation via corporate social responsibility: consumer priorities and the mediating role of food labels, *Agriculture and Human Values*, **33** (3), 597-609.

Draper, J., Dawson, M. and Casey, E. (2011) An exploratory study of the importance of sustainable practices in the meeting and convention site selection process, *Journal of Convention & Event Tourism*, **12** (3), 153-178).

Dunne, J., Chambers, K., Giombolini, K. and Schlegel, S. (2011) What does 'local' mean in the grocery store? Multiplicity in food retailers' perspectives on sourcing and marketing local foods, *Renewable Agriculture and Food Systems*, **26** (1), 46-59.

Edwards, D., Foley, C. and Hergessel, A. (2016) *Conferences: Catalysts for thriving economies*, report prepared for Business Events Sydney, UTS, Sydney. https://www.besydney.com.au/media/1611/catalysts-for-thriving-economies-research.pdf

Engelseth, P. and Hogset, H. (2016) *Adapting Supply Chain Management for Local Foods Logistics. Proceedings in System Dynamics and Innovation in Food Networks*, Innsbruck, Austria: International Center for Food Chain and Network Research, University of Bonn, Germany, pp. 143-160.

Enjolras, G. and Aubert, M. (2018) Short food supply chains and the issue of sustainability: A case study of French fruit producers, *International Journal of Retail & Distribution Management*, **46** (2). https://doi.org/10.1108/IJRDM-08-2016-0132

Estrada-Flores, S. and Larsen, K. (2010) *Best Practice Food Distribution Systems*, Melbourne: Food Chain Intelligence and Victorian Eco-Innovation Lab.

Feagan, R., Morris, D. and Krug, K. (2004) Niagara region farmers' markets: local food systems and sustainability considerations, *Local Environment*, **9** (3), 235-254. https://doi.org/10.1080/1354983042000219351

Foley, C., Edwards, D., Harrison, B. and Hergesell, A. (2017) *ICC Sydney: Feeding your performance*, report prepared for Joint Meetings Industry Council, Australia: UTS. https://www.themeetingsindustry.org/

Foley, C., Edwards, D., Harrison, B. and Hergesell, A. (2019) *ICC Sydney: Feeding Your Performance*. University of Technology Sydney and ICC Sydney. https://www.themeetingsindustry.org/web/wp-content/uploads/JMIC_Case-Study_ICC-Sydney_Feeding-your-performance_Y2_18.9.pdf

Foley, C., Edwards, D., Harrison, B. and Hergesell, A. (2018a) *Feeding your Performance program: Evaluation one year on*. report prepared for ICC Sydney, Australia: UTS. https://www.iccsydney.com.au/.

Foley, C., Edwards, D. and Harrison, B. (2018b) A case study in collaborative supplier partnerships, in J. Liburd and D. Edwards (eds), *Collaboration for Sustainable Tourism Development* (p. 225), Oxford, UK: Goodfellow Publishers.

Freeman, R. and Liedtka, J. (1991) Corporate social responsibility: a critical approach, *Business Horizons*, **34** (4), 92-98. https://doi.org/10.1016/0007-6813(91)90012-K

Gold, S. and Heikkurinen, P. (2013) Corporate responsibility, supply chain management and strategy, *Journal of Global Responsibility*, **4** (2), 276-291. https://doi.org/10.1108/JGR-10-2012-0025.

Harrison, B., Foley, C., Edwards, D. and Donaghy, G. (2019) Outcomes and challenges of an international convention centre's local procurement strategy, *Tourism Management*, **75**, 328-339.

Hemphill, T. (1997) Legislating corporate social responsibility, *Business Horizons*, **40**, 53-58. https://doi.org/10.1016/S0007-6813(97)90009-4

Hjalager, A. (2018) Suppliers as key collaborators for sustainable tourism, in D. Liburd and D. Edwards, *Collaborations for Sustainable Tourism Development*', Oxford, UK: Goodfellow.

Hughes, D.W., Brown, C., Miller, S. and McConnell, T. (2008) Evaluating the economic impact of farmers' markets using an opportunity cost framework, *Journal of Agricultural and Applied Economics*, **40** (1), 253-265.

IBISWorld (2017) Dairy Cattle Farming in Australia, viewed 15 March 2017. https://www.ibisworld.com.au/industry-trends/market-research-reports/agriculture-forestry-fishing/agriculture/dairy-cattle-farming.html

ICC Sydney (2016) *ICC Sydney Feeding Your Performance*, video file, viewed 15 March 2017. https://www.youtube.com/watch?v=EabNOVg0uiw

International Congress and Convention Association (2020), *2015 ICCA Statistics Report: Public Abstract*, ICCA, Netherlands.

Ilbery, B. and Maye, D. (2005) Food supply chains and sustainability: evidence from specialist food producers in the Scottish/English borders, *Land Use Policy*, **22** (4), 331-344.

Jackson, T, Zammit, K. and Hatfield-Dodds, S. (2020) *Snapshot of Australian Agriculture 2020*, Australian Bureau of Agricultural and Resource Economics and Sciences, Canberra. https://doi.org/10.25814/5e3a4ad8f80e7

Jarzębowski, S., Bourlakis, M. and Bezat-Jarzębowska, A. (2020) Short food supply chains as local and sustainable systems, *Sustainability*, **12** (11), 4715.

Jung, S., Kim, Y., Malek, K. and Lee, W. (2016) Engaging attendees in environmental sustainability at trade shows: attendees' perceptions and willingness to participate, *Anatolia*, **27** (4), 540-542.

Kim, Y. (2017) Consumer responses to the food industry's proactive and passive environmental CSR factoring in price as CSR tradeoff, *Journal of Business Ethics*, **140**, 307-321. https://doi.org/10.1007/s10551-015-2671-8

King, R., Hand, M. and Gómez, M. (eds) (2015) *Growing Local: Case Studies on Local Food Supply Chains*, Lincoln: University of Nebraska Press.

Kneafsey, M., Venn, L., Schmutz, U., Balázs, B., Trenchard, L., Eyden-Wood, T., Bos, E., Sutton, G. and Blackett, M. (2013) Short food supply chains and local food systems in the EU: a state of play of their socio-economic characteristics, *JRC Scientific and Policy Reports*, Seville: Joint Research Centre, Institute for Prospective Technological Studies, European Commission.

Kottila, M. and Rönni, P. (2008), Collaboration and trust in two organic food chains, *British Food Journal*, **110**, 376-394. https://doi.org/10.1108/00070700810868915

Krause, D., Handfield, R. and Tyler, B. (2007) The relationships between supplier development, commitment, social capital accumulation and performance improvement, *Journal of Operations Management*, **25** (2), 528-545.

Kumar, R., Agrawal, R. and Sharma, V. (2013) e-Applications in Indian agri-food supply chain: relationship among enablers, *Global Business Review*, **14** (4), 711-727.

Lee, W., Barber, T. and Tyrrell, T. (2013) Green attendees' evaluation of green attributes at the convention centre: using importance–performance analysis, *Anatolia*, **24** (2), 221-240.

Lee, S. and Slocum, S. (2015) Understanding the role of local food in the meeting industry: An exploratory study of meeting planners' perception of local food in sustainable meeting planning, *Journal of Convention & Event Tourism*, **16** (1), 45-60).

Lehtinen, U. (2012) Sustainability and local food procurement: a case study of Finnish public catering, *British Food Journal*, **114** (8), 1053–1071. https://doi.org/10.1108/00070701211252048

Leire, C. and Mont, O. (2009) 'The implementation of socially responsible purchasing', *Corporate Social Responsibility & Environmental Management*, **17** (1), 27-39. https://doi.org/10.1002/csr.198

Liburd, J. (2018) Understanding collaboration and Sustainable tourism development, in J.J. Liburd & D. Edwards (eds), *Collaboration for Sustainable Tourism Development*, Oxford, UK: Goodfellow Publishers, pp. 8-35.

Liburd, J. and Edwards, D. (eds), *Collaboration for Sustainable Tourism Development*, Oxford, UK: Goodfellow Publishers

Maloni, M. and Brown, M. (2006) Corporate social responsibility in the supply chain: An application in the food industry, *Journal of Business Ethics*, **68** (1), 35-52. https://doi.org/10.1007/s10551-006-9038-0.

Manning, L. (2013) Corporate and consumer social responsibility in the food supply chain, *British Food Journal*, **115** (1), 9-29. https://doi.org/10.1108/00070701311289858

Marsden, T., Banks, J. and Bristow, G. (2000). Food supply chain approaches: exploring their role in rural development. *Sociologia ruralis*, **40** (4), 424-438.

Mundler, P. and Laughrea, S. (2016)'The contributions of short food supply chains to territorial development: A study of three Quebec territories, *Journal of Rural Studies*, **45**, 218-229.

Myung, E. (2017) An exploratory study of attendee willingness to pay for environmentally friendly meetings, *International Journal of Hospitality & Tourism Administration*, **18** (3), 272-287.

News.com.au (2016) New ICC sommelier is up to the job choosing the 200,000 wines that will be poured each year, viewed 15 March 2017. http://www.news.com.au/national/nsw-act/new-icc-sommelier-is-up-to-the-job-choosing-the-200000-wines-that-will-be-poured-each-year/news-story/091e15c8e2cca84750e35a88214ea364

NSW Government (2014) NSW Waste Avoidance and Resource Recovery Strategy 2014-2021, viewed 18 April 2017. http://www.epa.nsw.gov.au/resources/wastestrategy/140876-WARR-strategy-14-21.pdf

NSW Government (2015) Economic Development Strategy for Regional NSW, viewed 18 April 2017. http://www.industry.nsw.gov.au/invest-in-nsw/news-and-events/news/new-economic-development-strategy-for-regional-nsw

Oglethorpe, D. and Heron, G. (2013) Testing the theory of constraints in UK local food supply chains, *International Journal of Operations & Production Management*, **33** (10), 1346–1367. https://doi.org/10.1108/IJOPM-05-2011-0192

Otto, D. and Varner, T. (2005) *Consumers, vendors, and the economic importance of Iowa Farmers' Markets: An economic impact survey analysis*, Leopold Center Pubs and Papers, Paper 145, viewed 15 April 2017. http://lib.dr.iastate.edu/cgi/viewcontent.cgi?article=1146&context=leopold_pubspapers

Porter, M. and Kramer, M. (2006) The link between competitive advantage and corporate social responsibility, *Harvard Business Review*, **84** (12), 78-92.

Pullman, M. and Dillard, J. (2010) Values based supply chain management and emergent organizational structures, *International Journal of Operations & Production Management*, **30** (7), 744-771. https://doi.org/10.1108/01443571011057326

Renting, H., Marsden, T. and Banks, J. (2003) Understanding alternative food networks: Exploring the role of short food supply chains in rural development, *Environment and Planning A*, **35** (3), 393-411.

Rimmington, M., Carlton Smith, J. and Hawkins, R. (2006) Corporate social responsibility and sustainable food procurement, *British Food Journal*, **108** (10), 824-837. https://doi.org/10.1108/00070700610702082.

Scarborough, P., Appleby, P., Mizdrak, A., Briggs, A., Travis, R., Bradbury, K. and Key, T. (2014) Dietary greenhouse gas emissions of meat-eaters, fish-eaters, vegetarians and vegans in the UK, *Climatic Change*, **125** (2), 179-192.

Shrivastava, P. (1995) The role of corporations in achieving ecological sustainability, *Academy of Management Review*, **20** (4), 936-960.

Shukla, M. and Jharkharia, S. (2013) Agri-fresh produce supply chain management: a state of the art literature review, *International Journal of Operations & Production Management*, **33** (2), 114-158.

Smith, N. (2003) *Corporate Social Responsibility: Not Whether, But How?*, Centre for Marketing Working Paper, **17** (1), pp. 1-35.

Sox, C., Benjamin, S., Carpenter, J. and Strick, S. (2013) An exploratory study of meeting planners and conference attendees' perceptions of sustainable issues in convention centers, *Journal of Convention & Event Tourism*, **14** (2), 144-161. DOI: 10.1080/15470148.2013.787662.

Spencer, S and Kneebone, M. (2012) *FOODmap: An analysis of the Australian food supply chain*, Canberra: Department of Agriculture, Fisheries and Forestry.

6

Springmann, M., Godfray, H., Rayner, M. and Scarborough, P. (2016) Analysis and valuation of the health and climate change co-benefits of dietary change, *Proceedings of the National Academy of Sciences*, **113** (15), 4146-4151.

Stoll-Kleemann, S. and O'Riordan, T. (2015) The sustainability challenges of our meat and dairy diets, *Environment: Science and Policy for Sustainable Development*, **57** (3), 34-48.

Székely, F. and Knirsch, M. (2005) Responsible leadership and corporate social responsibility: metrics for sustainable performance, *European Management Journal*, **23** (6), 628–647. https://doi.org/10.1016/J.EMJ.2005.10.009

Tikkanen, I. (2014) Procurement and consumption of local and organic food in the catering of a rural town, *British Food Journal*, **116** (3), 419–430. https://doi.org/10.1108/BFJ-05-2012-0116.

Tourism Australia (2017) Food and Wine, viewed 15 April 2017. http://www.tourism.australia.com/campaigns/Food-Wine.aspx

UNWTO (2015) United Nations declares 2017 as the International Year of Sustainable Tourism for Development, viewed 7 April 2017. http://media.unwto.org/press-release/2015-12-07/united-nations-declares-2017-international-year-sustainable-tourism-develop

UNWTO (2017) 2017 International Year of Sustainable Tourism for Development, viewed 30 March 2017. http://www2.unwto.org/tourism4development2017

Wiese, A. and Toporowski, W (2013) CSR failures in food supply chains – an agency perspective, *British Food Journal*, **115** (1), 92–107.

Yawar, S. A. and Seuring, S. (2018) The role of supplier development in managing social and societal issues in supply chains, *Journal of Cleaner Production*, 182, 227–237.

Zorzini, M., Hendry, L. C., Huq, F. A. and Stevenson, M. (2015) 'Socially responsible sourcing: reviewing the literature and its use of theory', *International Journal of Operations & Production Management*, **35** (1), 60-109.

# 7 Academic Scholars and Conferences

> *Academic events are powerful platforms that can be instrumental for the academic sector in fulling its role in society. The events allow for engagement with other sectors in society – these are platforms for dissemination of research; they are platforms for engaging stakeholders such as patient groups, media and industry; and they are key arenas for forming the networks and ideas that will shape the research of tomorrow.*
>
> Thomas Trøst Hansen, Ph.D., Science Attaché & Consul, Innovation Centre Denmark, Shanghai Royal Danish Consulate General

Ever since higher education became formalised, academic scholars have come together physically to share information, learn from each other and showcase their achievements. What most likely began in the Middle Ages as informal gatherings of like-minded thinkers had evolved into semi-structured symposiums (small conferences) by the 15th century, organised by scientific societies. These early conferences provided platforms for members to speak freely about subjects of mutual interest, exchange ideas and stimulate thought (Hankins, 2009; Sopta, 2013; Guerado, 2017; Rogers, 2019). For example, the great scholar Galileo was able to freely share his scientific discoveries with peers at the Italian Accademia dei Lincei (formed in 1603) despite his ideas being deemed extremely controversial and condemned by the church (Egger & Carpi, 2009). The knowledge and ideas generated and exchanged at scientific society symposia in the early modern period (15th to the 18th century) have been credited with making significant contributions to the development of scientific thought (Rogers, 2019).

Over time, these symposiums became more formalised, evolving into the larger conferences linked to the professional associations that we know today, where attendees have opportunities to network, form friendships and further their careers (Foley *et al.*, 2014). When viewed from this historical perspective, it becomes clear that academic research conferences are the foundational pillar upon which today's meetings and events industry has been built.

This long history, along with the unique traditions, conditions and expectations associated with academic conferences, presents a challenge for today's business event professional, who is eager to engage with academics in the bidding and hosting process and use their skills to increase the number, size and successfulness of such events, yet who has scant information or insight into the motivations, or cultural norms and practices of the world of academia. Why would a time-poor professor want to spend dozens of unpaid hours organising a conference? And what happens if the event is a flop? What's the point of a young researcher going along to a conference if no-one is interested in talking to them? Why don't more academics want to host conferences? How is a mid-career academic supposed to gain the skills or access the resources needed to organise their first major event? How does it feel when a PhD student spends months preparing a presentation and no-one turns up to hear it? And what are the real benefits of conferences for the universities and institutions that predominantly fund them?

Whilst academic conferences have been held for centuries, the conference industry which has been built around them is relatively young, only becoming professionalised in the latter part of the 20th century (Rogers, 2013). In the meantime, the global business events industry has grown exponentially, with air transport developments allowing mass travel in the second half of the 20th century, in turn prompting the formation of international communities and friendships among scholars who meet, often annually, to share findings and common interests (Foley *et al.*, 2014).

In the 21st century, technological advances in the area of communications and a health pandemic accelerated the growth of virtual participation in academic conferences. As issues such as the carbon footprint generated by large academic conferences, the increased capacity of new technologies to connect people who are geographically distanced, university funding cuts, and the travel restrictions experienced as a result of the recent health crisis arise (along with the potential for similar crises in the future), questions and challenges are emerging around the long-term viability of the traditional conference: in particular, do the potential benefits and outcomes of the current model warrant the costs and inputs required? It is crucial, therefore, to the future of academia, the knowledge economy and the events industry that fuller understanding and recognition of the value of academic conferences is attained, and that practical tools and systems are developed to ensure legacy outcomes are maximised.

This chapter focuses on some of the benefits, barriers and perceived outcomes of conferences for academics and universities. It specifically considers:

- ☐ The potential benefits of conference attendance for academics;
- ☐ The potential benefits of hosting conferences for academics;
- ☐ The potential benefits to universities as a result of their academics and PhDs attending or hosting conferences;

- The real and perceived barriers that discourage academics from hosting conferences;
- The relevance of these barriers and benefits to event professionals and suggestions for industry application.

The chapter includes a case study adapted from Hansen (2018), providing insights into the experiences of a number of academics who had each hosted (or attempted to host) an academic conference. This research makes an important contribution to our understanding of the benefits and barriers encountered by academics who undertake such a project and offers suggestions for how such benefits might be maximised and barriers overcome.

Following the case study is a more general discussion of the benefits for academics of attending conferences in the context of the credibility cycle (Hansen *et al.*, 2020). There is also discussion of how academics' involvement with conferences can benefit the broader community and examples of some significant outcomes which have already been achieved. Whether or not an academic conference is being organised by an individual academic, a professional association or event professionals, the majority of delegates, presenters and keynote speakers will be academics. As key stakeholders, it is vital that their needs and perspectives are well understood and considered in the delivery of the event.

## Case study: Benefits and challenges of hosting conferences for academics and universities

| | |
|---|---|
| **Business event(s)** | Six international research conferences and two failed attempts to host a conference in four European countries |
| **Location(s) and time of events** | Edinburgh (Scotland), Amsterdam (Netherlands), Gothenburg (Sweden), Copenhagen, Aalborg & Odense (Denmark), held between 2012 & 2016. |
| **Type of event(s)** | Academic conferences from various fields, including engineering, health, economics and biology |
| **Primary audience** | Various – with delegate numbers of between 458 and 20,522 |
| **Primary event legacy (for host)** | Increased visibility and networking capacity for academics and researchers organising international conferences |
| **Additional legacies (beyond tourism)** | *For host institutions* (universities): improved global reputation, attraction of talent, increased visibility, external funding<br>*For wider and local research environments*: increased access to top talent, increased visibility for researchers and their work, greater access to recent research for project development and teaching |
| **Barriers (for host)** | Lack of time and resources |
| **Summary quote** | "The conference itself is only part of a larger process from the motivation to host, to planning and execution to the follow-up phase. All four phases can spur benefits (for) the host institution, the research environments and for individual researchers." |

This case study was adapted from: Hansen, T. (2018)

7

## Introduction

Academics generally enjoy significant professional independence, however there are certain tasks which most are required to do. Teaching, for example, is a common priority for academics, and universities are placing increased emphasis upon research and publication outputs. However, when it comes to the importance institutions place upon encouraging academics to host research conferences, there is much less consistency or clarity. Whilst there is general consensus that hosting successful research conferences can raise the profile and networking capacity of the host researcher and the university, the human and (sometimes) financial resources invested into attracting, planning and delivering the event can be significant, and can impact an academic's performance of other important role requirements.

This case study aims to partially address this issue by highlighting the benefits associated with hosting academic research conferences. It also identifies barriers commonly experienced by those willing to embark on such an undertaking and suggests strategies to minimise or eliminate them. Data from eight planned European research conferences (six of which went ahead and two of which did not) is used to identify the benefits organisers perceived were derived from the events and the barriers they faced in delivery. The events were held/planned to be held in Scotland, The Netherlands, Denmark and Sweden between 2012 and 2016 with delegate numbers of between 458 and 20,522.

## The Danish meetings industry

For some years, the Danish meetings industry and related tourism organisations have highlighted the economic benefits derived from business meetings (VisitDenmark, 2012). Conferences and business meetings contribute to the host economy by way of job creation and tourism-related interactions with, for example, transport, accommodation, venue hire, catering, leisure facilities and the service sector at large (Grado *et al.*, 1997; Hanly, 2012; Lee *et al.*, 2013; Foley *et al.*, 2014) These direct economic contributions help to justify the resource investment where convention bureaux or peak tourism and destination marketing bodies are responsible for delivery of the event. However, it is a different story when responsibility for such an event falls into the lap of one or two academic staff. In fact, a significant number of the conferences held around the world each year are academic events, which presupposes the involvement of the local research environment in the whole event cycle, from attracting the event to executing and evaluating it.

In Denmark, the Ministry of Business and Growth is responsible for the continuing development of tourism, including the meetings industry. One of the main concerns for the future development of the industry is the continued ability to recruit and engage local university researchers in the bidding and hosting process of large events. To address this concern, they formed an advisory body to the Minister for Higher Education and Science and the Danish Council for Research and Innovation Policy, as part of the national tourism strategy launched in 2016.

## International research conferences in the Danish context

This case study includes information presented in two previous reports published in 2017 by the Danish Council for Research and Innovation Policy:

(1) *International Scientific Conferences: Marketplaces for Knowledge*

(2) *Case Analysis of International Research Conferences*

Danish consulting firm, IRIS Group, which specialises in business research and innovation policy, prepared the latter report which details six international conferences and two non-conferences (i.e. events that were planned, but not executed). These reports found significant evidence that universities and individual researchers who hosted research conferences were often placed firmly at the centre of attention in the international research environment. This then had positive impacts upon individual researchers, the research environment and the broader institution. On the other hand, they found that hosting large international research conferences required a strong commitment and significant resources from the host institution and, in particular, from the host researcher(s).

Bidding for, planning and executing such events can take several years and comprise a range of challenges at every stage. The local organising researchers often rely on external support to address these challenges. With this in mind, the reports investigated whether it was more beneficial for Danish academics to 'free-ride' by attending existing conferences organised by others, or to become hosts. They also considered whether hosting conferences was worth the effort in terms of increased academic quality and research capacity, and what barriers there are to hosting conferences.

## Impacts of scholarly work (including conference hosting) on the knowledge economy

The issues raised in the above reports are directly relevant to two current debates in the academic sector. They are:

- The importance of scholarly work to the knowledge economy;
- The importance and impacts of international academic research collaborations.

The first issue relates to the growing interest across the research and innovation sector about how to assess and communicate the diverse impacts of scholarly work. The notion of 'research impact' has gained importance in recent years, and has, to a varying extent, been embedded in research policies referring to universities' contributions to the knowledge economy (Benneworth *et al.*, 2016; Greenhalgh *et al.*, 2016; Reale *et al.*, 2017). The research impact agenda is also reflected in current research and innovation policies, where the linear model of high-quality scientific research trickling down to society is no longer seen as being adequate. However, until now, the importance of actually *hosting* academic events has been absent from this debate. These reports by the Danish Council of Research and Innovation Policy hint, however, at the relevance of including them in future impact assessment debates.

7

The second issue relates to the impact of international research collaborations on the quality, productivity and outcomes of research and the role of conferences in facilitating such collaborations. Whilst science policy practitioners agree that internationalising research is the pathway to increased quality and productivity (Lee & Bozeman, 2005; Wagner, 2008; European Commission, 2011; Sugimoto *et al.*, 2017), there are few policies or strategies in place to facilitate such internationalisation in Denmark.

Large research conferences, by their very nature, encourage sharing of knowledge and creation of collaborative networks amongst delegates from across the globe. It would follow, therefore, that such events would make a significant contribution toward the internationalisation of research. Yet few institutions have policies in place to encourage hosting of such an event in order to bring about such collaborations. In fact, of the eight universities in Denmark, not one refers to hosting research conferences as an objective of their internationalisation strategies.

This case study seeks to contribute to these debates by identifying the benefits derived and the barriers faced by academics involved in the organisation and delivery of international research conferences.

## *Impacts previously identified in the literature*

For the purpose of this study, the body of academic literature on the impacts of academic (research) conferences has been divided into two major categories (studies focused on *academic* impacts and studies focused on *socio-economic* impacts) and two sub-categories (*formal* impacts and *informal* impacts). These distinctions are shown in Table 7.1.

**Table 7.1:** The impacts of academic conferences

| | Academic impact | Socio-economic impact |
|---|---|---|
| Formal | Citations, publications, awards | Tourism, income to businesses and hotels, increased employment |
| Informal | Network creation, greater visibility, increased interdisciplinary funding, career advancement | Knowledge transfer to industry |

### Formal academic impacts

Academic impact studies generally utilise a bibliometric/scientometric approach, e.g. measuring citations and publications from conference proceedings, priors etc. Interesting impacts included:

- A 2011 analysis on the 'computer vision' subfield of computer science, which shows that journal papers created on the basis of conference 'priors' are significantly more cited than those without priors (Eckmann *et al.*, 2012).
- A bibliometric study from 2008, which shows the relative importance of conference proceedings has been diminishing over time. However, they still play a vital role in research progression in the fields of engineering and computer science, accounting for close to 20% of references (Lisée *et al.*, 2008). In other fields they're losing ground to other types of literature.

### Formal socio-economic impacts

These studies are typical in tourism research and are based on economic geography and econometrics. They focus on the socio-economic benefits of hosting conferences, e.g. the economic impact on growth and employment resulting from delegate expenditures on hotels, venues etc. From this perspective, conferences are generally viewed as being worth the costs of hosting them. For example, according to VisitDenmark, daily expenditure for international research delegates in Denmark is approximately 3100 DKK ($US491). Thus, these delegates comprise the segment within tourism that spends the most, fuelling intense international competition amongst tourism-related organisations to attract conferences (VisitDenmark, 2012).

### Informal academic impacts

Another strand of literature focuses on the derived or indirect research-related value added from conferences in terms of generating international research networks and gaining international exposure (Foley *et al.*, 2013; Foley *et al.*, 2014; Edwards *et al.*, 2017). These studies generally utilise sociological or anthropological approaches, focusing on social networks. Common examples include discussion of concepts such as:

- The positive impact of bringing different constituents to the same physical event in relation to learning, innovation and the development of a research field (Schüßler *et al.*, 2015);
- The capacity for conferences to benefit career development for elite scholars, organising committee chairs or members, program committee chairs or members, session chairs or invited speakers (Jeong *et al.*, 2009):
- The social role of conferences in entrenching existing power relations and hierarchies in research communities (Da Silveira *et al.*, 2015)

### Informal socio-economic impacts

This area of the literature includes complex, multi-layered research that considers how international conferences result in networks and knowledge sharing between academia and industry (Francisco *et al.* 2011; Chen, 2021). The literature on research conferences has investigated a number of perspectives on both socioeconomic and academic benefits of conferences. However, literature dealing with the value added from specifically *hosting* conferences is sparse.

## Analytical framework

The analytical framework of this analysis is based on several distinctions made in relation to:

1. The process;
2. The actors;
3. The benefits and barriers identified.

## *The conference process*

The conference itself is only part of a larger process stretching from the motivation and decision to host a conference, through planning and execution, to the follow-up phase. All four phases can, in theory, spur benefits for the host institution, the research environment and individual researchers, as outlined below.

- The very first step in the conference process is the initial decision to host. It is important, therefore, to consider the motivations or drivers behind that decision. One common driver is the direct relevance a particular conference has to the host, their research environment or their home institution. For example, the host can be driven by a desire to fill a gap in existing knowledge in their field, to be at the centre of the international network and/or to host a conference that draws elite global researchers.
- The next stage involves gaining approvals to host (from supervisors and/or the home institution), planning, bidding and lobbying (in the international research sector). This includes programming the conference concept, allocating resources and facilities, mobilising the international research community, calling for abstracts and the peer reviewing of papers prior to the conference.
- The third phase is execution of the conference, which can include every-thing from logistics (hotels, infrastructure and bureaucracy related to the influx of hundreds or thousands of international delegates), to overseeing keynotes, break-out sessions, networking activities and external communications.
- Finally, there are the follow-up activities aimed at ensuring embedding and continued momentum in the research results and networks generated at the conference. This includes publication of papers and conference proceedings, establishment of virtual communities and post-conference evaluations and research.

## *The actors*

The analysis focuses on the perceived benefits and barriers of hosting academic conferences from the perspectives of:

- The individual researcher
- The research environment
- The host institution.

Having 'ownership' of a conference is pivotal in understanding and obtaining the potential benefits such an event can bring. However, it is important to note that at large conferences such ownership will often rest with a scientific association or society which is seeking to extend networks and knowledge exchange between delegates from different countries or environments. The extent to which the conference 'owners' are engaged in organising the conference may also differ significantly. In many cases, scientific societies put conferences to tender among competing hosts or cities.

### *Benefits and barriers*

Benefits can be generated at all phases of the conference process for:

1. The host institution (increased collaboration with elite research environments abroad, highlighting of the research strengths of the institution, attraction of external funding etc.);
2. The research environment (enhanced research quality, increased internationalisation, coherence between researchers, talent development, educational development, new research agreements etc.);
3. Individual researchers (enhanced research leadership, better international standing in the global research community, publications in prestigious journals, new personal contacts, and career development because of increased visibility).

There can also be spill-over benefits for other environments at the institution as a result of new relationships and exposure to recent research outputs.

Some of the barriers to hosting conferences, which can make the task highly challenging and, in some cases, discourage academics and institutions from proceeding altogether, include:

- ☐ Lack of time and resources vis-à-vis core activities (research and education);
- ☐ Unclear goals or cost-benefits of hosting;
- ☐ Low prioritisation of conference hosting by the institution or from the surrounding research environment;
- ☐ Lack of experience or project management skills of host researcher and knowledge of practical and logistical issues.

Overcoming such barriers to generating benefits from hosting conferences can, to some extent, be mitigated by conference supportive framework conditions and instruments. These include support from convention bureaux (CVB), which might scout for conferences and host researchers, and support researchers in everything from the bidding process to the execution of the conference. Support staff at the host institution and/or professional conference organisers might also support researchers in the process, thereby minimising barriers.

## Research methods

The case study is based on semi-structured interviews with key informants involved in six conferences and in the planning of two conferences which did not proceed. The conferences were held (or planned to be held) between 2012 and 2016 in six different cities. The methodological approaches used were:

- ☐ Primary data (semi-structured interviews with key stakeholders);
- ☐ Secondary data (literature review).

## *Selecting the conferences*

The cases were selected in close dialogue with the Danish Council for Research and Innovation Policy and MeetDenmark, the national agency for attracting conferences. Three of the cases are Danish, while the remaining three are from Scotland, the Netherlands and Sweden. From the outset, it was important that the cases reflected a variety of conference profiles, based on the criteria below.

- Varied locations – in well-connected metropolitan areas and regional centres with capacity for hosting large conferences;
- Varied institutional profiles – including both research-intense, national 'Humboldtian' universities and regional universities ;
- Varied scientific topics – from areas such as health and humanities;
- Various sizes – from conferences with several thousand delegates (with high project management demands) to ones with a few hundred delegates;
- Various periods of time having elapsed since the event occurred – to provide some long-term perspectives on outcomes and to monitor the impacts of fading memory;
- Inclusion of events which did not proceed – to gather information on barriers.

**Table 7.2:** Characteristics of the eight conferences included in the study

| Conference | Location | Scientific topic | Approx. size |
|---|---|---|---|
| 31st International Conference on Thermoelectrics, 2012 | Aalborg | Engineering | 600 |
| 12th International Family Nursing Conference, 2015 | Odense | Health care | 458 |
| European Society for Medical Oncology Congress, 2016 | Copenhagen | Medicine | 20,522 |
| 11th International Congress on the Biology of the Fish, 2014 | Edinburgh | Natural science | 485 |
| 51st Congress of the ERA-EDTA (European Renal Association and European Dialysis and Transplant Association), 2014 | Amsterdam | Medicine | 8146 |
| 28th annual congress of the European Economic Association and 67th European Meeting of the Econometric Society, 2013 | Gothenburg | Economics | 1656 |
| Non-case 1 European Social Science History Conference 2020 | Copenhagen | History | 1000-1500 |
| Non-case 2 International Society of Biomechanics Conference 2019 | Copenhagen | Natural science | 1000 |

## *Conducting the interviews*

Thirty-one semi-structured interviews were conducted with a range of stakeholders, including;

- Lead researchers;

- ☐ The scientific societies behind the conferences;
- ☐ Representatives of the institutional leadership;
- ☐ Representatives from local convention bureaus, pcos, and/or the venues at which conferences were held.

The interviews examined the following topics:

- ☐ Background and motivation for hosting the conference;
- ☐ Initial activities before execution of conference;
- ☐ Execution;
- ☐ Follow-up activities;
- ☐ Value added benefits and impacts from hosting;
- ☐ Barriers, challenges, their causes as well as solutions;
- ☐ Context and critical conditions for effects;
- ☐ Conference-promoting framework conditions, instruments and partnerships.

The qualitative analysis explored commonalities and determined the extent to which findings could be generalised. It identified critical factors and their significance for researchers and host institutions when bidding for international conferences. It also examined the importance of critical preconditions and contextual factors, such as the interplay between the university and the research environment, as well as the importance of experience and competencies in hosting conferences. This knowledge is crucial to developing an understanding of the causes leading to an observed benefit. Process tracing was used to combine interview responses with knowledge on impacts, benefits, barriers and causes to provide an in-depth, 'thick' description of causes of observed impacts or barriers.

## Discussion of findings



The findings are divided into three sections. The first section sums up the findings related to the individual host researcher. The second section addresses how universities and their local research environments benefit from hosting conferences. In the final section the barriers that are often encountered when hosting such events are discussed.

### *Benefits for the host researcher*

This section identifies and discusses the various benefits derived from hosting an international research conference from the perspective of the host researcher.

#### Visibility

There is little doubt the various tasks involved in planning, promoting and then delivering a research conference will increase the visibility of any academic taking on the role. This increased visibility is of most value in terms of raising the academic's profile in their respective international research community. As the central point of the conference, the host is able to interact with (and meet) research colleagues from all over the world. All host researchers stated the conference presented them with a unique opportunity to engage in networks and informal discussions about the newest research and potential future research collaboration. The researchers pointed out that physical attendance simply makes it easier to arrange impactful meetings with the international elite.

Much of the interaction took place outside the official program, such as during coffee breaks, in the hotel lobbies, and even before and after the conference. In particular, life science conferences attract delegates before and after the official conference program because they want to spend time in the local research environment, visit hospitals and meet PhD students. The host is often a central focus during a conference. They often deliver a welcome speech and usually introduce a presidential address or keynote speech by an acknowledged researcher. These events are usually very well attended. The host researcher also has more information about the program, venue, city etc. and normally welcomes and spends more time with the most renowned visiting researchers. One host believed the number of citations likely to be made of their work in the years following the conference would increase as a result of higher visibility.

## Career progression

While the most important factor for advancing a research career is high impact research, researchers whose visibility is high (perhaps as a result of hosting a conference) tend to be recognised in the research environment. Some interviewees noted that leading a successful conference does not in itself pave the way for promotion at the university, but suggested those most likely to host a conference are often ambitious, driven and passionate about the local research environment, hence more likely to get ahead regardless. The Congress and Programme Chair of the Biology of the Fish Congress said researchers who engage in conference bidding are usually appointed deans or are at a similar position later in their career.

Interview responses indicated most host researchers are actively engaged and known in the international scientific societies associated with their field prior to bidding. Thus, hosting an international research conference can potentially open doors for advancing within scientific societies. This is especially the case for life science conferences. The host researcher at the European Society for Medical Oncology (ESMO) Congress said hosting the conference was probably an important step towards a new position in the international scientific society.

## Speaking invitations

Several interviewees said they had been invited to speak at other conferences, run specific workshops and teach at summer schools since hosting, although they could not definitively say the invitations were as a result of their hosting role alone. Many are very active senior researchers, with reputations that may have warranted such invitations, however they did feel acting as conference host kept them 'on the radar' of the research environment and scientific societies.

### Improved organisational skills

In all six cases, the host was to some extent involved in the practical planning and organising of the event. The scientific societies often played a major role in the practical execution of the conferences, but some of the host researchers said their organisational skills improved as a result of hosting the conference. Several of the PhDs who were involved in conference hosting also noted this benefit, stating the process brought them out of their normal 'comfort zone', and provided them with improved management skills.

An example is the European Economic Association (EEA) Econometric Society Meeting (ESEM) in Gothenburg, where eight PhD students were involved in different tasks. Two were heavily involved, organising and coordinating 25 masters student volunteers. It was full-time work for two weeks before and during the conference. The masters students were paid a salary by the scientific society, while the PhD students could claim their hours as teaching time. Even though the PhD students did not attend many sessions because of their responsibilities they said they enjoyed the role immensely and acquired significant organisational experience which enriched their PhD experience.

## *Benefits for the host university and research environment*

This section identifies and discusses the various benefits derived from hosting an international research conference from the perspective of the host university and the ways in which the local research environment derives benefit.

### Exposure and promotion

The most commonly noted benefit of conference hosting from the perspective of the host university/research environment is the exposure and promotion generated. Whilst the direct benefit of such exposure is difficult to measure, researchers, university managers and convention bureaux representatives interviewed were clear the benefits were significant. They included:

- Invitations to join in new research consortia (two cases);
- Interaction between young researchers from the environment and experienced researchers from the global elite (one case);
- Increased investment from industry (two cases);
- More researchers interested in the research area (one case).

These advantages are more pronounced if the conference topics are directly relevant to host university research specialties. Where conference programs were organised by scientific societies (rather than the host university) host researchers managed to include special sessions relevant to the institution's research agenda, as illustrated by the following comment:

> *"The added value of hosting, is the opportunity to influence the academic program so that it set the spot light on some of the key research areas for the school,"*
> (Host researcher, University of Gothenburg).

7

Furthermore, it was found the added value of exposure and promotion differs in relation to the importance placed upon such exposure in the early planning stages and the strategies implemented. For example, the host of the 2016 ESMO congress actively used the congress to showcase Denmark's cancer research and treatment strengths and leverage benefit, thanks to implementation of a targeted communication strategy.

### Ability to attract new research talent

Interviews with organisers of three of the six conferences included in this study said it was easier to recruit top researchers to the university/research environment in the years following the event. This was viewed as an indirect effect of exposure to the international research environment. Delegates could explore the university and host city, and picture themselves living and working there. Several convention bureau staff noted this benefit for the host university was common, based on feedback from previous conference hosts.

### Access to new funding opportunities

Obtaining commitments for future funding is an important aspect of research conference networking. This was evident across all cases studied, with researchers explaining that getting together in informal meetings is an essential part of committing to new research projects requiring joint support and/or funding. The host researcher in Edinburgh stressed that personal meetings were essential to gain commitment to multimillion-euro research projects, and that many research funds (e.g. Horizon 2020) require involvement of an international consortium of researchers to be successful in funding applications. International conferences are important agoras (markets) to form such consortia as commitment does not happen via virtual communication alone.

The attendance of industry at conferences is also important for establishing new consortia for research projects. This is more relevant to natural science conferences. Like international collaboration, inclusion of industrial partners is often a requirement of applications for research funding. Both of the life science conferences in this study included extensive industry-sponsored symposia and room to display new products, technologies, etc.

Hosts of the 2016 EEA–ESEM event in Gothenburg also said that since holding the conference it had been frequently used to generate funds. The host school is currently fundraising for visiting professorships among local firms in Gothenburg, and references to the success of the conference was often more tangible and relevant for non-academic funding partners than name-dropping top journals in which one had published articles.

### Development of PhDs and young researchers

It is customary to invite PhD and occasionally Masters students from the host university to assist the researcher plan and execute the event. Students gain free access to the conference and are in some cases paid a small salary. Interviewees noted students had the opportunity to meet top researchers, network and, in some cases, present their own research to an international audience. This opportunity is a great supplement to doctoral education which most students would otherwise not be able to afford. In Edinburgh, for example, the host researcher encouraged a local PhD student to organise a symposium, and in Aalborg a PhD student presented their research results, which were subsequently published in an international journal.

### Greater research capacity

In all six cases, the main purpose of the conference (for both hosts and delegates) was for it to act as an agora for research improvement. In some cases, the host university formed a group of researchers who were assigned different conference tasks (i.e. one would evaluate abstracts, one would identify conference themes, etc.). These groups have continued working together, adding value to the local research environment.

When big scientific societies control conferences, the potential to strengthen local research environments is somewhat reduced, however local delegates will still benefit from visibility and exposure to the latest quality research, which in turn will benefit the research environment. In two of the three Danish cases, for example, the number of articles submitted to the conference by Danish researchers was higher than when the conference took place abroad. This means that even conferences organised predominantly by scientific societies can generate more publications and citations for researchers at the host university and in the national research environment.

### Stronger internal networks

In two of the six cases, interviewees from the university explained that a big event like the conference strengthened both the internal network of the hosting department and the local research environment. Researchers worked together to prepare and execute the conference and students played an active role, thereby getting to know their teachers.

> *"The conference was great for the internal relationship at the department. It was a unique opportunity for the students to meet the professors in another setting"*
> (Project Manager, University of Gothenburg).



### Income

The management of revenue or deficit varies between conferences and it can be an economic upside for the host university if the conference generates a surplus. In some instances, a distribution key is present; in other cases, the scientific society receives all surpluses. An example of a distribution key is the one used by the EEA–ESEM annual congress where the two scientific societies (The European Economic Association and The Econometric Society) and the host university equally share any potential surplus from the conference. Potential economic benefit, however, was not mentioned as a driver to host the conference by any interviewees. On the contrary, in most cases the host institution incurs considerable costs in staging the conference.

## *Barriers to hosting conferences*

There are also barriers (challenges) to hosting conferences. These can deter academics from considering the undertaking and/or discourage them from proceeding from one stage to the next. This section explores those challenges and discusses strategies used by hosts to overcome them. While challenges primarily constitute nuisances in an otherwise well-functioning process, barriers in the context of large research conferences being organised by one or a few researchers (such as during the bidding phase) are more critical and can prevent the conference from occurring. The barriers are analysed in terms of the significance they held in decisions over whether or not to proceed and the prevalence (occurrence) across cases studied (see Table 7.5).

**Table 7.5:** Barriers and challenges

| Barriers and challenges | Significance | Prevalence across cases |
|---|---|---|
| Time and resources | High | High |
| Challenges in the bidding process | High | Medium |
| Conferences given low priority by home institution or department | High | Medium |
| Economic risk of hosting and lack of deficit guarantees | High | Medium |
| Handling value added tax | Medium | Medium |
| Limited influence on conferences organised by scientific societies | Medium | Medium |
| Lack of coherence in institutional goals and the broader institution | Medium | Medium |
| Weak follow-up measures | Medium | High |
| Regulations on academia-industry interaction at life science conferences | Medium | Low |
| Challenges in motivating young researchers | Low | Medium |
| Organisational challenges | Low | Medium |

## Time and resources

In the pre-conference phase, one of the most widespread and significant barriers reported was lack of time and resources on the part of the researcher – a barrier which is highlighted in four of the six cases where host researchers were required to commit more than 300 hours to organisation of the conference. Hosting conferences is generally a residual activity for researchers that demands that they either take time away from their obligations (research and/or teaching) or use their own personal time. When added to the potential risks (failures in event management, loss of personal prestige if things go wrong, economic deficits etc.), the transaction costs of hosting can appear insurmountable for individual researchers. Furthermore, researchers are primarily recognised, merited, and receive funding based on time spent on core research and educational activities, which means the incentives for organising conferences are comparatively much weaker.

*"In academia, it is not possible to de-prioritise research or educational obligations. At the same time, there is a high degree of self-management, so I have been able to plan myself. But it involved a great deal of work outside of my fixed working hours to make ends meet"* (Host researcher, Aalborg University).

*"The trade-off between time spent on research and conference planning can be an issue. In general, it is well balanced at the school. But a conference will always take someone's time that could have been spent on research"* (Vice-Dean of The School of Business, Economics and Law, University of Gothenburg).

However, this barrier is not as significant where conferences are organised primarily by the host institution or scientific society. Some medical conferences studied, for example, were funded, organised and executed by well-resourced professional associations with comparatively fewer resources needed from the host institution.

## Challenges in the bidding process

The bidding process itself can be challenging, with winning bids needing to demonstrate a convincing value proposition vis-à-vis competitors (i.e. on the attractiveness of the location and venue, the host's ability to organise the conference, suggestions for set-up and events, budget etc.). In some cases, the convention bureau handles some of these tasks. Participants indicated researchers can lack a number of competencies in the bidding phase, including:

- Lobbying skills and influence over key stakeholders (i.e. at the scientific societies);
- Ability to construct a convincing value proposition (compelling writing/marketing skills);
- Ability to understand and address the formal requirements of the bidding process;
- Competencies with regard to preparing and planning a conference program;
- Knowledge in identifying/conveying what the host city offers;
- Skills in budget preparation to demonstrate financial viability of the conference.

Local convention bureaux (Wonderful Copenhagen, Convention Edinburgh and Amsterdam Marketing) supported researchers in the bidding process for some of the conferences studied. This support included helping researchers understand the formal requirements of a conference tender, providing input on the value offerings of the city and assisting with writing skills and graphic layout so as to prepare a convincing value proposition. These agencies also supported bids by funding presentation trips by the researcher (e.g. Convention Edinburgh) and/or planning site visits for scientific society representatives.

However, interview results indicated there was little support available from convention bureaux in terms of direct lobbying efforts. This is because the bureau would have little knowledge of the internal politics or processes of the scientific society or which key stakeholders to influence. For example, in the case of the International Congress on the Biology of the Fish in Edinburgh, a bid was lost because the organising society had an internal rule of thumb that only every third conference could be held outside North America and the previous conference had been held in Europe.

## Low prioritisation of conferences by home institution or department

Lack of support from departmental or university management was noted as another major barrier. Compared to research and education, the organisation of conferences appears not to be generally recognised as a priority by management, despite the potential benefits. While the institutional managers interviewed said they welcomed conferences that were of high quality and could contribute to "putting the university on the map", they generally viewed organisation of such events as a residual activity, in line with other outreach activities. Thus, with few exceptions, institutional support for those organising conferences appears to be relatively ad hoc. This perceived inconsistency can confuse and discourage researchers considering whether or not to bid for an event, and was the reason one planned conference did not go ahead.

The lack of management support fell under two broad areas:

- ☐ A lack of encouragement or support – where there was little institutional backing (morally or practically) in the bidding or preparatory phase;
- ☐ A lack of post-event support – where there was no recognition or merit given for hosting the event.

Both conferences which failed to go ahead noted lack of institutional support as the reason. A host at the University of Copenhagen made the observation:

> *"Apart from an initial positive attitude from the dean there was no consistent management backing in the bidding process, nor economic support in shape of a deficit guarantee or the like. It was a one-man show, with very large risks as I did not know how the attendance would turn out. This happened during a time of large budget cuts and layoffs at the university, which made it way too risky for us – so it was in a way a relief to lose the bid [...] If I should do it again, management backing would be absolutely crucial".*

Another interviewee explained the lack of support this way:

> *"A university is like a layered cake. The university management often does not know much about what the researchers are doing and vice versa. They are generally not heavily involved in conferences"* (Manager Meetings & Conventions at Amsterdam Marketing).

## Economic risk

The economic risk of hosting a conference (e.g. as a result of low attendance, unforeseen costs etc.) can be discouraging, according to several interviewees. This is particularly important in cases where the researchers bear a large share of the organisational responsibility for the conference vis-à-vis the scientific societies. While it is important to conduct a thorough assessment of risks vs. costs, the potential insecurity may impede researchers and the institution from bidding in the first place. In the case of the Biology of the Fish conference (Edinburgh), the local organiser was not granted a deficit guarantee. While this posed a risk, the researcher also noted it forced him to seek out good deals at local hotels and suppliers in close collaboration with the convention bureau.

## Handling taxes

How to manage VAT (namely which expenses can be deducted and how) posed a challenge for half the hosts interviewed. Interviewees from convention bureaux noted the challenges relate to various income and expenditure items, such as venue rentals, delegate fees, meals and travel expenses. It is a complicated system and is particularly relevant in countries such as Denmark and the Netherlands. Some challenges identified were:

- ☐ The judicial status of the lead organiser ;

- ☐ The extent to which the 'conference owner' (e.g. the professional association) makes a profit ;
- ☐ How to handle industrial income (e.g. from exhibitors);
- ☐ How to develop clear and transparent contracts between conference partners addressing vat responsibility.

### Limited influence on conferences organised by societies

Whilst society-driven conferences demand less input of researcher or host institution resources and assume most of the risk, they also offer fewer potential benefits (motivators) for the host. In four of the cases, the international scientific society played a very extensive role in planning and executing the conference. The most prominent example is the large annual congress of the ERA-EDTA, which generally does not draw in local universities in the preparatory or execution phase. This also means the broader institutional effects of hosting are very limited. The main local beneficiaries of hosting were the two local members of the organising committee.

### Weak follow-up

It was concerning to note that in all cases there was little done by way of knowledge-gathering for the future, developmental evaluation, or sharing of information about the host experience. While most of the cases undertook some knowledge gathering via surveys, little was done to systematically embed any networks developed or reap long-term benefits from the experience. Some of the challenges reported, include:

- ☐ **A general lack of long-term strategy or clear goals for hosting**. When the ex-ante goals of hosting, e.g. to increase international visibility, attract funding etc., are unclear it becomes difficult to make a targeted ex-post follow-up.
- ☐ **Lack of institutional involvement**. In most cases studied, the conference had limited connection to the host institution outside the immediate research environment. Thus, many researchers expressed disappointment that. given the conference was not imbedded in the host institution's/department's activities, leadership quickly forgot what it accomplished, reducing opportunities for ongoing impacts at the institutional level.
- ☐ **Follow-up conducted internally by societies**. In society-driven conferences, the follow-up benefits primarily lie within the society, while the local hosts' involvement in the follow-up can be limited.

### Regulations limiting academia-industry interaction at conferences

Participants from two large medical conferences and two convention bureaux raised the issue of compliance with international regulations regarding interaction between industry and healthcare professionals at conferences. These regulations seek to balance the need for knowledge exchange between industry and the healthcare sector with concerns about conflicts of interest and industrial influence over health care. However, the regulations also create a barrier for organisers via limiting potential for joint activities, e.g. symposia.

Interviewees noted two primary challenges concerning knowledge and awareness of these frameworks:

- Frequent rule changes (creating difficulties for host researchers and advisers)
- Varied interpretations across countries, institutions and/or public authorities).

### Challenges in motivating young researchers

Young researchers were encouraged to focus attention on producing quality publications to progress their research careers. Interviewees said this made it more challenging to motivate them to participate in conference planning.

> *"Young researchers are very oriented towards their career and generating high-impact articles. They seem more interested in fulfilling their ambition and career - and often do not have the time"* (Business Development Manager, RAI Amsterdam).

### Organisational challenges

Most researchers emphasised that organising a major event such as a research conference required extensive practical skills and competencies. Whilst most of these challenges were eventually overcome, they were (at times) stressful. Some interviewees were originally unaware of where to go for assistance, but once they connected with local convention bureaux or societies, assistance was provided. Some of the practical challenges bureaux were not able to help with, however, were:

- Forming and leading a competent, dedicated local planning and execution team;
- Identifying good keynote speakers, given that relevant keynotes typically require a thorough understanding of the international research environment.

## Conclusions

In summary, this case study highlighted several benefits related to hosting academic events. For the individual researcher, the benefits were related to visibility, career advancement, invitations to speak at other events and improved organisational skills. The most prevalent and common benefit was visibility. Further research is needed to understand relationships between these benefits (i.e. how does visibility impact career advancement or invitations to speak?). It would also be of benefit to investigate benefits by career stage (i.e. do senior researchers experience different benefits to junior researchers?).

For host institutions, the benefits were related to exposure and promotion of the research environment, recruitment of top researchers, access to application consortia, education of PhD students, quality of research, strengthened internal networks and income for the host university. Of these, the most significant and prevalent were benefits related to promotion of the institution and recruitment of new talent. Similarly to the researcher-related benefits, it was unclear how the benefits were interrelated, nor how they impacted the long-term reputation, research and teaching outputs, and overall performance of the institution.

Furthermore, the case study highlighted several barriers inhibiting the realisation of these benefits. Of these, the most significant was a lack of time and resources to carry out the necessary tasks involved in hosting the conference in the first place. Some barriers can be overcome by collaborating with meeting industry actors, including professional conference organisers and larger overarching societies or industry groups.

A next step for research in this area could be to explore under which circumstances the potential benefits of hosting such events become great enough to overcome barriers. The study does not offer an answer to the initial policy-driven question of whether hosting conferences is an optimal strategy for researchers compared to free-riding by participating in conferences in other places. In order to explore this question, a theoretical model of how benefits underpin each other, and which are most important, could be developed. Moreover, it would be worth considering whether benefits and barriers vary by specific scientific discipline, type or size of meetings, and host region. Such research might assist the industry to recruit and engage local university researchers in the bidding for and hosting of conferences.

## Theoretical discussion

As demonstrated in the case study above, academics and institutions are able to benefit from a broad range of potential legacies as a result of hosting research conferences. In particular, those academics who organise and host such events can raise their professional visibility and networking capacity, which generally results in enhanced career progression. The host university can also increase its regional and international visibility and reputation, its capacity to attract global talent and, subsequently, its attractiveness to prospective students. Hosting academic conferences also enables universities to provide their early-career researchers with opportunities to network and present their research to an international audience.



However, for such beneficial outcomes to be fully realised, organisers and event industry practitioners need to know what is driving the key stakeholders – the academics themselves. They need to discover *why* academics attend (or don't attend) conferences; they need to know *what* it is that motivates academics to present their work at conferences; and they need to understand *how* hosting an academic conference can create long-term career benefits for those willing to embark upon the task.

Once such understanding is gained, conference delivery strategies can be developed that will ensure those goals are attained. To illustrate, the leading scientists in any discipline are the ones who earn their colleagues' recognition (Garcia & Sanz-Menendez, 2004) and conferences offer the perfect opportunity for such recognition to be obtained. If it is clear, then, that visibility and recognition are key outcomes sought by academics in return for the time and effort spent on hosting a conference, event professionals might build opportunities to showcase the academic's achievements into the event program or to 'lionise' them in media releases, publicity or promotions.

As Hansen *et al.* (2020) state, the meetings industry needs to understand its role as part of the system of knowledge production and utilisation. Event professionals also need to genuinely confront the barriers that make event hosting daunting for many academics, such as by providing additional administrative support, offering deficit guarantees or assisting with technology and promotion. Essentially, they need to create the necessary tools, processes and systems to ensure barriers are overcome, benefits are maximised and outcomes are measured and documented.

Beyond identifying some of the benefits academics can derive from hosting conferences, the case study also raises some related questions for both industry and academia. For example, how do these benefits relate to, or depend upon, each other; do all academics leverage these benefits, and how well prepared are our early-career researchers to leverage the benefits available to them from conferences? Finally, given that many of those interviewed cited a lack of support, reward or recognition for hosting conferences from their own university as a very real barrier to their willingness to become involved, is this rational in terms of the many benefits which the institution can gain? Or, if the potential benefits are *not* being recognised, attained or valued by the institutions, why not?

## Interdependencies between conference benefits

Recognition from peers is currency for academics and conferences have a well-established reputation for providing platforms for academics to present work, argue, and gain recognition (Hansen *et al.*, 2020). Bourdieu (1975) makes the point that academics are driven by need for recognition of their command of their field of research in order to achieve social capital and exercise social power. A second interesting point is that this recognition is largely bestowed on academics by the academic peers in their field (Bourdieu 1975). Academic conferences provide an important mechanism through which academics are able to make personal and career gains in terms of recognition from other academics. This is true for all academic delegates and perhaps even more so for hosting academics.

Hansen *et al.* (2020) refer to recognition as credibility and provide a model of the ways in which academics invest in a process of building credibility, including at conferences. The various forms of credibility to be gained via conferences are presented as buzz, networks, grants, equipment and data, arguments, presentations, scholarly output and recognition (see Figure 7.1). Buzz, for example, refers to the process of getting a feel for the issues and trends currently impacting their field. It is an essential aspect of capital and credibility for academics who wish to be at the forefront of developments in the field. It can include both formal and informal activities or conversations which may occur at conferences. Those connecting with the buzz are often seen keenly interacting with others, networking or raising relevant issues or questions during poster sessions, and they are clearly inspired and energised by all that is occurring around them. Delegates whom one might refer to as being really 'into' a conference are connecting with this 'buzz' and (intentionally or unintentionally) building credibility as a result. Interestingly, those most likely to be 'into' a conference are also those who are really 'into' their field. However, to build their careers academics also require other forms of credibility such as grants and scholarly output.

The process of gaining academic credibility is not linear. Not only can various forms of academic credibility be achieved at different times, but they can also impact each other. Many interdependencies have been found. For example, as illustrated in Figure 7.1, findings from a study of specialist conferences (Hansen *et al.*, 2013, p. 485) indicate that networks are vitally important to gaining credibility and that they also impact (and are impacted by) other forms of credibility. Indeed, conferences are ideal settings for academics to establish and strengthen networks. Foley *et al.* (2014) argue that the 'time out' aspect of conferences contributes to this. They note that for the few days of the conference academics are usually somewhat removed from the stresses and distractions of everyday life, which gives them time to engage meaningfully with like-minded peers from around the world. Hansen *et al.* (2020) found that networking helped academics to access equipment and data, as well as grants, and recognition was shown to be helpful in forming networks. Further research is highly likely to find many more connections between the various forms of credibility for academic conferences.

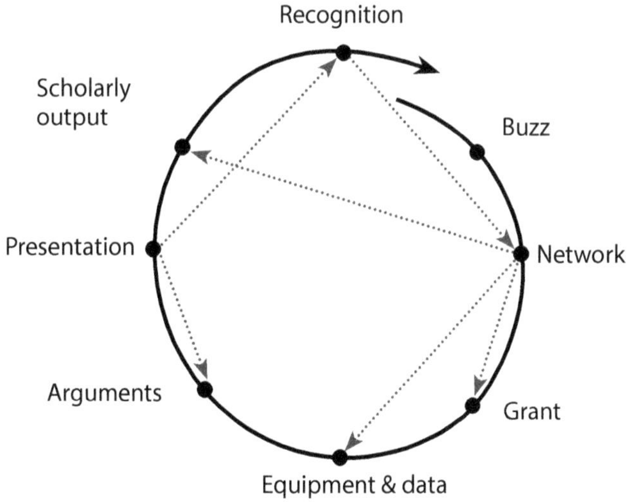


**Figure 7.1:** Academic credibility cycle. (Source: Hansen *et al.*, 2020, p. 485)

Conferences are hugely important for many, if not all, of the forms of credibility needed to support the careers of academics and their endeavours. But more importantly, achieving the various forms of credibility at conferences allows academics to generate broader impacts that bring benefits to communities around the world. Indeed, most universities have a mission that involves bringing value to global communities. For example, the University of Technology Sydney (2021) states "we will be defined by our support for the economic, social and cultural prosperity of our communities... and we are committed to research, innovation and the dissemination of knowledge of public value". So how are forms of academic credibility at conferences linked to outcomes for communities?

The AIDS conferences of the late twentieth century provide a useful illustration. The HIV/AIDS epidemic emerged as a major international health crisis in the early 1980s. The disease was rampant in many countries, particularly in less-developed regions. The academics and clinicians operating in those areas were relatively isolated and, at times, ill-informed as there was no central mechanism for coordinating the knowledge and experiences those in the field were accumulating. Political, cultural and inter-agency divides meant establishment of a formal central body to deal with the issue was problematic, however organising a conference was less of a challenge. The international AIDS conferences enabled people to come together to compare notes, experiences and understandings (see Foley *et al.*, 2010; Foley *et al.*, 2013; Edwards *et al.*, 2017). The buzz at the AIDS conferences gave delegates access to the most relevant and pressing issues associated with the disease and the presentations provided cutting-edge scientific and medical research and explanations about what was going on. The buzz, networking, and arguments between delegates from various backgrounds helped to break down cultural barriers which had precluded some from discussing pertinent and sensitive issues (for example, around sex, homosexuality, drug injection) that were related to infection control. The conferences helped academics to network, share findings and gain access to valuable information and data, ideas, techniques and technology. All of the forms of credibility played a part in supporting academics to help the global community to manage the epidemic more effectively.

In another example, Professor Tom Calma AO and The Honourable Linda Burney MP credit the buzz at the Australian Aboriginal Education conferences of the 1990s with driving progressive social policy that saw the doubling of both the number of Indigenous students enrolled in tertiary education and Year 12 retention rates in ensuing decades (Edwards *et al.*, 2017). The conferences enabled Aboriginal leaders and experts to have a say in the education of both Indigenous and non-Indigenous children and paved the way for the improved outcomes. Calma was involved with those early conferences, and says they gave people from a diverse range of backgrounds the opportunity to hear about the issues that mattered. He believes they were one of the key drivers behind the social and political shifts that have occurred for Indigenous Australians in the last few decades. According to Burney, the value of the conferences lay in the way they enabled organisers to bring different people with different perspectives together into the buzz. "It's as simple as that: they put people together. Certainly, there was a very definite political element to the conferences and participants were always mindful their overall objective was to bring about policy change, but it was the way the conferences brought everyone together to establish common goals that helped us achieve the level of success which we did" (Burney, cited in Edwards *et al.* 2017, p. 117). Both Burney and Calma credit the buzz at these conferences with setting an agenda for change.

Despite such accounts, and the potential value of the buzz for progressing beneficial outcomes and developing careers, it is the least researched of the forms of credibility. This is most likely because grants, presentations and scholarly output, for example, are far more tangible and easily measured. However, buzz abounds at most conferences and is available to all delegates. Good keynote speakers often provide their view of the issues and developments in the field to kick-start the buzz, and these topics can be at the forefront of panel discussions and many formal presentations. Buzz also abounds in less formal settings via the many discussions that take place between sessions, during lunch and tea breaks, and at the various social events that make up the conference program.

## The capacity of academics to leverage conference benefits

However, understanding whether all academics (including hosting academics) actively access buzz and other forms of credibility is hard to determine as very little research has been done in this space. A growing body of literature on academics and conference legacies can, however, give us an indication.

Whilst some academics are well aware of the drivers, benefits and motivations for hosting and attending academic conferences, there are also many who are unaware or unclear about how the experience can be used to progress their research or career.

Edwards *et al.* (2017) investigated the conference practices of high-achieving Australians and cite many examples of academics who networked at conferences to gain funding for projects, partners for research and ideas for future research direction. For example, Nobel Prize recipient and leading astrophysicist, Professor Brian Schmidt, credits the networks and relationships he built at conferences as crucial to his research success (Edwards *et al.*, 2017). Schmidt and his team established that our universe was actually expanding, not shrinking (as had previously been assumed); a discovery which has likely saved billions of dollars in wasted research hours and failed experiments. Similarly, it was a chance conversation at a conference with a like-minded peer which launched Professor Ian Frazer on the path to develop the HPV vaccine, which has significantly reduced the risk of cervical cancer for women around the world (Edwards *et al.*, 2017).



However, Wang *et al.* (2020) investigated the conference experiences of early career academics (including PhD students) and got quite a different result, finding some academic delegates do not leverage much value at all from conference attendance. Indeed, many did not attend social events and sessions beyond their own paper presentation session, which meant they did not have access to buzz or opportunities to network. Reasons for non-attendance were varied and included having no interest in the topics, having competing priorities (work, leisure, family commitments), feeling physically and mentally exhausted, and experiencing feelings of social exclusion. This was particularly the case for some PhD students who found the conference experience overwhelming, alienating and/or irrelevant (Wang *et al.*, 2020). These findings suggest that many PhD students were aware of the value of conferences as a platform to present their research but were less aware of the potential for conferences to assist with other forms of credibility. This is an area that may benefit from further research.

Further, whilst academics at all stages of their career (early, mid or later career) can benefit from the opportunities conferences offer, research suggests the dominant drivers can be different at different stages of one's career. For example, one study found that early-career researchers are most interested in having opportunities to attain new knowledge and establish beneficial connections at conferences, whilst those in the later stages of their careers are more likely to be interested in reaffirming connections with those already within their networks or in opportunities for more personal interactions with those they know and trust (Edwards *et al.*, 2016).

Wang *et al.* (2020) found many early-career researchers attended conferences seeking opportunities to learn from or create research connections with middle and later-career academics, and were more inclined to have meetings with (or attend sessions given by) those delegates they perceived to have social influence, rather than spend time on presentations delivered by other PhDs, whom they saw as being relatively poor in terms of social capital wealth. Wang *et al.* (2020) also found many early-career researchers, in particular PhD students, found the conference environment to be unexciting, unsupportive and uninviting. They complained that some conference communities were 'closed', making it hard to join conversations. Other theorists have noted that, even if conference planners consciously create an environment that attempts to include students in social events, sometimes it simply does not work (Látková *et al.*, 2009). Wang *et al.* (2020) reference the ideas of Portes (2000) to explain the exclusion from the group that prompts some PhD students and early-career academics to stop attending conference sessions. They argue that the social capital and trust relationships which are built up through social connections (such as the networks later-career academics have developed with each other over the years of conference attendance) can create a shared network of resources through which they are able to promote their interests. However, this can also have negative effects, including encouraging the exclusion of outsiders and repressing the development of members. The actions which discourage outsiders from entering the community and accessing the coveted resources may be covert and carried out unconsciously by member actors at the conference (Portes, 2000). Regardless, these actions are still likely to result in outsiders being isolated as a result of established formal and informal structural barriers and, as such, it is essential that strategies of social inclusion are employed at conferences to make all feel welcome (Wang *et al.*, 2020).

## Potential benefits of conference hosting for universities

As illustrated in the case study, universities which host academic conferences benefit in a number of ways. By hosting a conference, the university is seen to be supportive of the field and at the forefront of developments. They are able to gain legitimacy and enhance their reputation as being relative and progressive. Higher education is a very competitive arena and reputation is one of the most significant factors used by students to select where they will study. More demand for a particular institution means the quality of student outcomes may also increase. This can have numerous flow-on impacts, including increased income, a higher quality of teaching talent, improved graduate outcomes and increased research funding. There is also exposure and promotion for the surrounding community as a result of tourism impact and this can increase the institution's social capital in the region.

The university's PhD students are usually given opportunities to present their research at a conference or help with its organisation, both of which improve their career prospects. Students gain exposure to the newest research in their field and the perspectives of leading academics. They are able to meet and form connections with other researchers at their level or with middle and late-career academics with whom they may one day partner in research or employment.

By hosting a conference, the university provides their whole teaching and research community (from both within the discipline and in other fields) with opportunities to meet and connect with peers from across the globe. As a result, connections and partnerships are formed, the university's assets are showcased and the local area can be explored. These things all combine to increase the likelihood that new talent will be attracted to the university in the future and that better research outcomes will be achieved. Once again, this can also contribute to greater social capital as the university is seen to be an employer capable of bringing new residents into the community and contributing to growth.



There is little doubt that networks and partnerships forged in the conference environment often result in funding partnerships. This is even more so for the central figures at a conference, including the host academic. The host is seen to hold significant status at many academic conferences, and has easy access to most attending delegates. The credibility that results often means a greater number of networks are established and the likelihood of opening funding negotiations increases.

The host of an academic conference often has direct access not only to all the presentations and papers, but also to those delivering them. Again, the networks and contacts formed often lead to successful research partnerships and increased likelihood of funding, particularly where grants can be applied for collaboratively. The outputs of these collaborative research partnerships, once published, again positively impact the reputation and promotion of the university.

Often those working in large institutions are isolated from those operating outside their immediate research area. As detailed in the case study, the process of hosting a conference, however, can see internal networks formed which may otherwise not have existed. These can be either interdisciplinary or cross-disciplinary. Such networks can again lead to research, teaching or operational collaborations which propel the university forward and maximise the potential of its staff. Such benefits will usually also impact teaching quality, student experiences and graduate outcomes.

Whilst there may not be a significant direct profit from the hosting of a conference (creation of profit is rarely a priority for academic conference organisers), there can be indirect economic benefits via increased student fees, sponsorships and grant opportunities. Some conferences may result in a small profit and there is usually an economic benefit to the local community through provision of goods, services and accommodation (Foley *et al.*, 2019). If leveraged strategically, income may be generated as a result of partnering with local service providers.

Interestingly, each of these benefits contributes to a continuous cycle of benefit. For example, the enhanced reputation attracts more students, which increases demand and income (from fees) and pushes up entry standards. Higher quality students usually result in better student outcomes, which in turn pushes up demand. The enhanced reputation, quality and exposure of students also enables the institution to attract new talent, which means better teaching and research output and, again, higher demand, better graduate outcomes and an enhanced reputation for the institution.

Despite the significant potential benefits to universities that host conferences, many of the academics interviewed for the case study cited the lack of support, recognition or reward offered by their employing university as one of the significant barriers they encountered in the process of hosting an academic conference. Perhaps it is time to ask why conference hosting is not seen by some in universities as a productive use of their academics' time. This issue has been raised by other commentators, who questioned why publication of prolific journal articles was seen as being an extremely important aspect of an academic's workload, yet conference hosting or presenting was not. As Rowe (2018) states,

> *"It is surely in the mutual interest of funders, organisers and the ASP [academic, scientific, professional] community to take positive steps towards making sure that the Cinderella status of conference presenters is finally ended and that going to the academic ball provides as much demonstrable and sustainable value as does writing an academic paper."*

# Lessons learned

There is a long history of academics attending conferences. The relationship of academics and conferences brings benefits not only to the academics themselves, but to their universities, their academic fields of scholarship and the global community as a whole. Academics use conferences to develop and expand networks, initiate collaborations, gain recognition, further their careers, build global communities of friendship with like-minded peers in their fields, learn about developments in the field, promote and disseminate their research and, ultimately, to make contributions to the broader community.

Hosting a conference has the potential to amplify these benefits for the hosting academic. However, most academics are time-pressed and hosting and attending conferences is time-consuming. It is important, then, to leverage the time invested in conferences towards the best possible outcomes.

Outcomes are often interdependent. Visibility as a conference host may support efficient networking, and showcasing interesting research may lead to an invitation to be part of a research grant team. However, outcomes are often dependent on the status of the academic and the effort they are willing to invest in the event. Hosting can provide additional access to social status, networks and ultimately recognition. Being involved in the delivery of a conference also has the potential to help PhD students with their career progression; however, many PhD students report less than exemplary experiences at conferences as they sometimes feel excluded from academic cliques of influence.

Beyond being seen simply as a means of generating tourism expenditure, it is important that event professionals, conference hosts, universities and academic delegates (including PhD students) understand the opportunities which conferences offer to:

- Develop stronger communities and outputs of research;
- Make significant contributions to the knowledge economy;
- Position universities as centres of thought and professional leadership;
- Enhance academics' careers.

It is also important that they develop and execute strategies which will enable measurement of such benefits and ensure they are not only achieved, but maximised.

The credibility cycle offers a useful model for understanding the ways in which conferences support the research and careers of individual academics. However, it is important to note that the forms of credibility – buzz, networking, grants, equipment and data, argument, presentations, scholarly output and recognition – can also underpin substantial legacies for global communities.

## References

Benneworth, P., Gulbrandsen, M. and Hazelkorn, E. (2016) *The Impact and Future of Arts and Humanities Research*, London: Palgrave Macmillan UK. https://doi.org/10.1057/978-1-137-40899-0

Bourdieu, P. (1975) The specificity of the scientific field and the social conditions of the progress of reason, *Social Science Information*, **14** (6), 19-47. doi: https://doi.org/10.1177/053901847501400602

Chen, W. (2021) Understanding the role of conferences for industry-university collaborations, PhD Thesis, UTS, Sydney.

Da Silveira, M., Bufrem, L. and Caregnato, S. (2015) Scientific events, power relationships and practices of researchers, *Transinformacao*, **27** (3), 199-208. https://doi.org/10.1590/0103-37862015000300002

Eckmann, M., Rocha, A. and Wainer, J. (2012) Relationship between high-quality journals and conferences in computer vision, *Scientometrics*. https://doi.org/10.1007/s11192-011-0527-2

Edwards, D., Foley, C. and Hergessel, A. (2016) *Conferences: Catalysts for thriving economies*, report prepared for Business Events Sydney, UTS, Sydney. https://www.besydney.com.au/media/1611/catalysts-for-thriving-economies-research.pdf

Edwards, D., Foley, C. and Malone, C. (2017) *The Power of Conferences: Stories of Serendipity, Innovation and Driving Social Change*, Sydney: UTS ePress.

Egger, A. E., & Carpi, A. (2009) Scientific Institutions and Societies. *Visionlearning* POS-2 (4),

European Commission (2011) Communication from the Commission to the European Parliament, the Council, the European Economic and Social Committee and the Committee of the Regions Horizon 2020: The Framework Programme for Research and Innovation.

Foley, C., Edwards, D. and Schlenker, K. (2014) Business events and friendship: Leveraging the sociable legacies, *Event Management*, **18** (1), 53-64.

Foley, C., Schlenker, K., Edwards, D. and Lewis-Smith, L. (2013) Determining business event legacies beyond the tourism spend: An Australian case study approach, *Event Management: An International Journal*, **17** (3), 311-322.

Foley, C., Schlenker, K., Edwards, D. and Hayllar, B. (2010) *A scoping study of business events: beyond tourism benefits*, report prepared for Business Events Sydney, UTS, Sydney. http://hdl.handle.net/10453/16848

Foley, C., Edwards, D. and Schlenker, K. (2014) Business events and friendship: Leveraging the sociable legacies, *Event Management*, **18** (1), 53-64.

Foley, C., Edwards, D. and Hergessel, A. (2019) *Direct expenditure attributed to international conferences held in New South Wales 2017-2018*, report prepared for Business Events Sydney, University of Technology Sydney. https://www.besydney.com.au/media/3360/bes002_de6_conferencesinfographic_a4-2.pdf

Francisco, M., Milojevic, S., Sabanovic, S. and Šabanovic, S. (2011) Conference models to bridge micro and macro studies of science, *Journal of Artificial Societies and Social Simulation*, **14** (4), 13.

Garcia, C. and Sanz-Menendez, L. (2004) *Competition for funding as an indicator of research* competitiveness: The Spanish R & D government funding*, University Carlos III, Madrid, Spain.

Grado, S., Strauss, C. and Lord, B. (1997) Economic impacts of conferences and conventions, *Journal of Convention & Exhibition Management*, **1** (1), 19-33. https://doi.org/10.1300/J143v01n01_03

Greenhalgh, T., Raftery, J., Hanney, S. and Glover, M. (2016) Research impact: a narrative review, *BMC Medicine*, **14** (78). https://doi.org/10.1186/s12916-016-0620-8

Guerado, E. (2017) Scientific societies and the third industrial revolution – the future role of the OTC. *Injury*, **48** (16), S1–S4.

Hanly, P. (2012) Measuring the economic contribution of the international association conference market: An Irish case study, *Tourism Management*, **33** (6), 1574-1582. https://doi.org/10.1016/j.tourman.2011.12.010

Hankins, J. (2009) Humanist Academies and the 'Platonic Academy of Florence.' In *Proceedings of the conference, 'From the Roman Academy to the Danish Academy in Rome'*, ed. R. Jensen and M. Pade. Analecta Romana Instituti Danici Supplementum. Copenhagen: Odense University Press.

Hansen, T. (2018) *International Research Conferences: The Academic Impact*, Aalborg University and Wonderful Copenhagen. https://www.themeetingsindustry.org/value-proposition/case-studies/

Hansen, T., Pedersen, D. and Foley, C. (2020) Academic events: an empirically grounded typology and their academic impact, *Event Management*, **24**, 481-497.

Jeong, S., Lee, S. and Kim, H. (2009) Are you an invited speaker? A bibliometric analysis of elite groups for scholarly events in bioinformatics, *Journal of the American Society for Information Science and Technology*, **60** (6), 1118–1131. https://doi.org/10.1002/asi.21056

Látková, P., Wu, H. C. J. and Paulsen, R. D. (2009). Examining social capital in the conference setting: A case study of college student participants. *SCHOLE: A Journal of Leisure Studies and Recreation Education*, **24**(1), 38-49.

Lee, C., Lee, M. and Yoon, SH (2013) Estimating the economic impact of Convention and Exhibition Businesses, using a regional Input–Output model, A Case Study of the Daejeon Convention Center in South Korea, *Asia Pacific Journal of Tourism Research*, **18** (4), 330-353. https://doi.org/10.1080/10941665.2012.658414

Lee, S. and Bozeman, B. (2005) The impact of research collaboration on scientific productivity, *Social Studies of Science*, **35** (5), 673-702. https://doi.org/10.1177/0306312705052359

Lisée, C., Larivière, V. and Archambault, É. (2008) Conference proceedings as a source of scientific information: A bibliometric analysis, *Journal of the American Society for Information Science and Technology*, **59** (11), 1776-1784. https://doi.org/10.1002/asi.20888

Portes, A. (2000) The two meanings of social capital, *Sociological Forum*, **15** (1), 1-12.

Reale, E., Avramov, D., Canhial, K., Donovan, C., Flecha, R., Holm, P., Larkin, C., Lepori, B., Mosoni-Fried, J., Oliver, E., Primeri, E., Puigvert, L., Scharnhorst, A., Schubert, A., Soler, M., Soòs, S., Sordé, T., Travis, C. & Van Horik, R. (2017) A review of literature on evaluating the scientific, social and political impact of social sciences and humanities research, *Research Evaluation*. https://doi.org/10.1093/reseval/rvx025

Rogers, T. (2013). *Conferences and Conventions: A global industry*, 3rd edition. London: Routledge.

7

Rogers, K. (2019) Biology, *Encyclopedia Britannica* (online), retrieved 30 December 2020. https://www.britannica.com/science/biology/Advances-to-the-20th-century#ref498750

Rowe, N. (2018) Conferences mean high times but low returns, *The World University Rankings*, retrieved 22 January 2021. https://www.timeshighereducation.com/features/conferences-mean-high-times-low-returns

Schüßler, E., Grabher, G. and Müller-Seitz, G. (2015) Field-configuring events: arenas for innovation and learning?, *Industry and Innovation*, **22** (3), 165-172. https://doi.org/10.1080/13662716.2015.1038098

Sopta, K. 2013. The Scientific Societies and Its Role in Creating a Socially Responsible Science. *Periodicum Biologorum*, **115** (1), https://hrcak.srce.hr/101235.

Sugimoto, C., Robinson-Garcia, N., Murray, D., Yegros-Yegros, A., Costas, R. and Larivière, V. (2017) Scientists have most impact when they're free to move, *Nature*, **550** (7674), 29-31. https://doi.org/10.1038/550029a

University of Technology Sydney (2021) Vision, accessed 10 January 2021. https://www.uts.edu.au/about/uts-2027-strategy/vision

VisitDenmark (2012) The economic contribution of meeting activity in Denmark. https://docplayer.net/21157526-The-economic-contribution-of-meeting-activity-in-denmark-published-by-visitdenmark-may-2012-isbn-978-87-87393-86-7.html

Wagner, S. (2008) *The New Invisible College*, Washington DC: Brookings Institution Press,

Wang R., Foley, C. and Edwards, D. (2020) It's quite common to skip sessions: A grounded theory approach to examining PhD students' non-attendance at academic conferences, in *20:20 Vision: New Perspectives on the Diversity of Hospitality, Tourism and Events*, CAUTHE 2020.

# 8 Business Events and Innovation

> *Most academic events would not be possible without the services provided by the meetings industry; the industry provides the essential infrastructure. In this sense, the meetings industry is comparable to laboratories, instruments and big scale infrastructure like CERN. A thriving academic sector depends on the services of meeting infrastructure. This understanding has allowed academics to consider the services provided by the meetings industry as a more legit service provider.*
>
> Thomas Trøst Hansen, Ph.D., Science Attaché & Consul, Innovation Centre Denmark, Shanghai Royal Danish Consulate General

As discussed throughout this text, there is growing awareness that business events create unique and powerful opportunities for collaboration and network formation (Jago & Deery, 2012; Veal *et al.*, 2012: Foley *et al.*, 2014; Jonson *et al.*, 2015; Edwards *et al.*, 2016; Crowther *et al.*, 2018). In addition, research on the long-term impacts of business events highlights the important role they play in knowledge creation and driving innovation (Foley *et al.*, 2010; Foley *et al*, 2013; Edwards *et al.*, 2016; Edwards *et al.*, 2017; Vlasov *et al.*, 2017; Crowther *et al.*, 2018). This emerging body of literature is important, contributing to a gap that was previously noted as "nonexistent in the current scientific landscape" (Hansen, 2010: 28). In this chapter we show ways in which business events serve as powerhouses of innovation, driven by unique aspects of their social spaces that bring together the experts in a field to gain and share knowledge, collaborate and network.

While acknowledging that there are countless definitions of innovation in the literature (literally thousands), for the purposes of this chapter we define it as the intentional generation and realisation of new ideas (or modification of old ideas to create new understandings) that suggest new ways to achieve objectives or apply new work methods. Innovation is a multistage process, beginning with the generation of a novel and useful idea to address a problem, then the promotion of that idea to potential allies (supporters) and, finally, production of the idea or product (perhaps via development of a prototype or model that can be realistically applied) (Ye *et al.*, 2021).

Startups play a key role in driving the process of innovation. A startup is a company, partnership or temporary organisation which has been formed to search for a repeatable and scalable business model by bringing new ideas to the market which are then transformed within economically sustainable enterprises (Spender *et al.*, 2017). Because the ideas which drive startups are new, they challenge the status quo and provide competition to the entrenched players that already hold positions of power within an industry. Without startups, and the entrepreneurs who drive them, many fields, such as technology, finance and healthcare, would stagnate, as long-standing players would continue to dominate using traditional methods and much slower internal innovative practices.

But by their very nature, startups are usually small, significantly under-resourced and inexperienced in business operation. In particular, they commonly lack the financial and human resources needed to see their innovative ideas through to the prototyping, testing, production and marketing stages of establishing a financially viable business (Spender, *et al.*, 2017). It is here that business events can play a role, not only in generating new ideas, but in connecting startups, the drivers of innovation, with larger industry players and public authorities seeking new ideas to provide a competitive edge, or to generate solutions to challenging problems.

The case study presented in this chapter focuses on the Swiss Fintech Corner, which was part of the 2016 Swift International Banking Operations Seminar (SIBOS), a premier business forum for the global financial community, held in Geneva, Switzerland. The case study assesses the value of strategically bidding for this business event that linked to the local knowledge and innovation network. It considers whether it raised Geneva's profile as a fintech hub, and also whether it increased awareness of the existence and importance of fintech startups amongst public and political authorities.

Against the background of this case study, the link between innovation, global knowledge networks and business events (both in general and more specifically within the business events context) will be discussed in the theoretical considerations section of this chapter. We will also touch briefly on the potential for the business event environment to be designed in ways which provide opportunities for play and experimentation; useful strategies in stimulating entrepreneurial and innovative thinking by attendees. Finally, we look at some of the challenges facing startups in today's competitive, dynamic environment, discuss the need for government policies that support startups and consider the potential for business events to drive innovation within industry sectors by connecting startups with larger industry players.

## Case study: Swiss Fintech Start-up Corner

| | |
|---|---|
| **Business event** | A dedicated start-up corner as part of a larger industry conference in 2016 |
| **Location(s)** | Geneva, Switzerland |
| **Types of event** | The Swiss Fintech Corner was part of the 2016 SIBOS (Swift International Banking Operations Seminar). |
| **Business event purpose** | The aims of the Swiss Fintech Corner were: to showcase Geneva (and Switzerland) as a hub of financial technology innovation; to enable invited startups to present their services to a broad local and international audience of venture capital companies. |
| **Primary event legacy** | Increased Switzerland's profile as a centre of innovative financial technology and an important financial marketplace. |
| **Additional legacies (beyond tourism)** | Raised profile of Geneva as a fintech hub. Increased awareness of fintech startups amongst public and political authorities. Connected local fintech entities with global financial institutions. Some increase to the profiles of Geneva and Zurich as important financial centres. |
| **Summary quote** | "The global financial community loved the idea because it allowed them to connect directly to smaller companies with bright ideas on how to tackle big issues for our industry." |

This case study (CS) was adapted from: Scaglione, M. and Dimitriou, S. (2018)

## The Swift International Banking Operations Seminar (SIBOS)

SIBOS is an annual conference, exhibition and networking event organised for the financial industry by the Society for Worldwide Interbank Financial Telecommunication (SWIFT), a global provider of secure financial messaging services. First held in 1978, the event has now evolved to become a premier business forum where those in the global financial community are able to debate the future of payments, securities, cash management, and trade and financial crime compliance. Each year it is hosted in a different location and brings together business leaders, decision makers and topic experts from financial institutions, market infrastructures, multinational corporations and technology partners. The 2016 SIBOS event in Geneva hosted 8300 attendees from 158 countries, and included hundreds of speakers and conference sessions, nearly 200 exhibitors and multiple networking events over a four-day program. The SIBOS organising team was headed by Sven Bossu, who worked closely with the local economic minister, Pierre Maudet, to bring leading bankers to the business event and facilitate introductions between local operators and experts, who were happy to share their knowledge. The overall aim of the conference was to generate discussion on business strategy and build networks in order to collectively and collaboratively shape the future of the financial industry. This business event was the first ever to incorporate a fintech (financial technology) startup corner.

## The Swiss Fintech Corner

The Swiss Fintech Corner was just one aspect of the wider conference. It was set up with the help of the Economic Development Office of Geneva, Alp ICT (a startup accelerator for technology enterprises in Switzerland) and the Swiss Finance & Technology Association. The aims of the Swiss Fintech Corner were:

- ☐ To highlight Switzerland and Geneva's capacity and potential for innovative fintech delivery;
- ☐ To create opportunities for local fintech entities to establish relationships and connections with larger financial industry players.

Of the 216 registered and relevant in Switzerland, the organising committee selected 24 to participate in the Swiss Fintech Corner. These 24 were considered to reflect the Swiss traditions of financial services excellence and technical innovation, and to most appropriately meet the specific selection criteria. That criteria required the startup to have:

- ☐ An up-and-running solution;
- ☐ A clear pricing-plan;
- ☐ A base in switzerland;
- ☐ A relevant solution.

(Swiss Finance & Technology Association, 2016).

Five cantons were represented by the startups and three of the five main sectors of activity were:

- ☐ Institutional investment;
- ☐ Banking infrastructure;
- ☐ Cryptofinance and blockchain.

**Figure 8.1:** Sibos layout (left) and Swiss Fintech Corner at the Swiss Lounge at Sibos.

## Research and theoretical context

A number of important contextual factors are relevant to the significance, interpretation and understanding of this case study.

### *The fintech sector*

Fintech became an important sector in its own right in the wake of the global financial crisis. Technological innovation created the opportunity for significant expansion of choice for financial services consumers (Deloitte, 2016:3). Instead of traditional financial services offering packaged products, new innovative fintech companies offer customised solutions, which are often seen as more efficient and cost-effective.

### *Switzerland and Geneva as fintech hubs*

Switzerland is one of the most important financial centres in the world, home to more than 325 banks and managing 27 per cent of transnational assets (startupticker.ch, 2016). Switzerland's status as a financial centre is integral to its national identity (Ernst & Young, 2016:3) and the importance of the industry for the national economy is evident, given it contributes directly to approximately 10 per cent of the national economy and 6 per cent of full-time equivalent jobs (BAK Basel Economics AG, 2015:3f).

When looking at Switzerland as a fintech hub, however, it has lagged in terms of start-up activity when compared to other innovation-driven hubs such as New York City, London and Singapore (Ernst & Young, 2016).

According to the Global Financial Centres Index, which is used to measure global financial centres in terms of their competitiveness, Switzerland's two financial centres, Zurich and Geneva, have both been losing ground over recent years. Zurich was down from 7th to 9th position in 2016 and Geneva dropped even more significantly from 13th to 23rd position (Z/Yen Group, 2016). Despite this, Switzerland overall has a number of important competitive advantages over other centres, which could see it easily recover lost ground. These are:

- Its reputation as an innovation hub;
- Its centralised geographic location;
- The high level of security in the country;
- Its reliability, quality and stability;
- Its large financial centre;
- Its excellent infrastructure, legal system and reputation for protecting consumers.

(Deloitte, 2016)

### *Geneva as a meetings industry destination*

Geneva's reputation as a meetings industry destination was another consideration which informed this case study. Since its re-organisation in 2014, the policy of the Geneva

Convention Bureau has focused on the promotion of business events belonging to local clusters of high added value competencies, as defined by the Economic Development Office (The Iceberg, 2017).

## Rationale of the research

The banking and finance industry is one of Geneva's excellence clusters. As a result, the opportunity to host the SIBOS conference and establish the Swiss Fintech Corner at the conference were the Geneva meeting industry's policy targets. The aim of this research is to determine what benefits were achieved as a result of these targets being met. More specifically, the research seeks to determine whether benefits (beyond tourism income) were achieved as a result of strategically bidding for business events which linked to local knowledge and innovation. In particular, it focuses on whether Geneva's position and visibility as a fintech innovation hub were enhanced, and whether the Swiss finance and banking sectors were seen to be more innovative as a result of the event. Finally, the case study seeks to identify the level to which the startup entities involved were able to significantly increase their contacts, networks and reach, and whether their audience base (customers and investors) was expanded internationally.

### *Research methods*

The case study is based on both primary and secondary data in the following forms:

- Semi-structured interviews with stakeholders and policy makers;
- A survey of the 24 start-up entities involved in the Swiss Fintech Corner;
- Basic demographic data on delegates attending the SIBOS business event;
- Social media (Twitter) metrics.

### Semi-structured interviews with major SIBOS stakeholders

The primary data was collected using semi-structured interviews guided by a questionnaire which focused primarily on the benefits for Geneva at the macroeconomic level. The interviews were conducted in March, 2017, on four people, who each represented one of the main stakeholder groups at the conference (details later in case study).

In the interview introduction, researchers made the point that the questionnaire was designed to investigate the impact of the Swiss Fintech Corner *only*; not the SIBOS conference itself. It was also stated that the research focus was on impacts *other than* direct economic ones and that the aim was to measure impacts on:

- The image of Geneva as a meeting, incentives and convention (MICE) destination;
- The fintech ecosystem in Geneva;
- The startups of the fintech sector.

Each interview took 30–45 minutes.

The interviews were structured around four questions, reproduced below.

**Question 1**: How do you think the fintech corner at SIBOS affected the visibility of Geneva as:

- a tourism destination in general?
- a meeting industry destination in particular?

**Question 2:** How does Geneva's reputation as an important finance/banking marketplace affect the visibility and/or synergies of the startups at the Swiss Fintech Corner in terms of:

- public opinion in general and the political authority's awareness in particular?
- related companies (such as banks, insurance, health, etc.)?
- education and research at the universities and applied university level?
- competitive position with Zurich as a fintech hub and competitive position at an international level (i.e. the Global Financial Centres Index ranking)?

**Question 3:** How do you think the start-up corner affected the role/involvement of traditional sectors (banking, insurance, health, others)? What about venture capital companies? Do you think the business event created new synergies between startups or increased their competition?

**Question 4:** Other comments and suggestions for improvements?

Interviews were collected and recorded by researchers via Skype and manually transcribed. The only exception was the interview with Sven Bossu – where SIBOS handled the recording through Cisco WebEX. Unfortunately, this recording failed. The interviewee then agreed to answer the interview questions again in writing. The researchers acknowledge the limitations in the methodology, but no alternative was possible within the scope of their project.

Interviews were analysed using Sphinx IQ2, Quali software, using a lexical and statistical approach. The lexical approach identifies the most frequently used words and expressions (i.e. Swiss Fintech Corner). One of the main goals was to show the homogeneity of opinions across the interviewees. The statistical approach helps to estimate the level of homogeneity across interviewees. To this end, cross-tabulations between the main words or expressions yielded by the lexical analysis and the interviews were calculated (i.e. principal components analysis, $\chi2$ test).

8

## *Start-up entity surveys*

To examine the benefits realised on a microeconomic level, a questionnaire capturing the satisfaction levels of the Swiss Fintech Corner startups was analysed. This questionnaire had been designed, distributed and collected by the Geneva Economic Development Office and Alp ICT immediately after the Swiss Fintech Corner ended, and was external to this research initiative. It was re-analysed for the purposes of this study. Sixteen of the 24 companies responded and the Geneva Economic Development Office provided the raw data for this study. The satisfaction survey among the startups was analysed using descriptive statistics.

### *Demographic data of SIBOS delegates*

The SIBOS survey (SIBOS, 2017) was administered by the conference organising committee and sought to establish demographic data on the 8300 delegates in attendance, rather than the Swiss Fintech Corner specifically. It showed that for 93% of those sampled, the conference met or exceeded their expectations. Forty-nine per cent of delegates were from Europe, Middle East and African regions, 24% were from the United Kingdom or the Nordics, 14% were from the Americas and 13% were from Asia and the Pacific. The main business focus represented in the four-day business event was payments. Other primary business focus areas included trade service, cash management and securities. Different institutions were involved in the business event, such as commercial banks and software vendors/consultancies. Finally, the survey showed the main reasons for attending were networking, keeping up to date with industry trends, conducting business and listening to debates.

### *Social media metrics*

The other source of secondary data was the social media communication metrics gathered during the exhibition and Swiss Fintech Corner, with specific attention to Twitter. The communications strategy was led conjointly by the Swiss Finance and Technology Association and Alp ICT. The metrics concern the tweets of the Swiss Finance and Technology Association and other parties posted during the Swiss Fintech Corner, using the hashtag #swissfintechcorner and belonging to the Twitter handle: https://twitter.com/finnovationCH.

The data yielded a time series of the evolution of the following metrics: new followers, tweets, total tweet impressions, link clicks, re-tweets, likes and average engagement rate. All the series were monthly. The series length was one year, from March 2016 to March 2017, except for the 'new followers' data where the data started from July 2016.

For the social media data structural time series was used (Koopman *et al.*, 2009). However, in most cases, the reduced length of the time series only allowed estimation using a classical ordinary least squares model. A model was built for each variable. All the models had the independent variable time (t) and the dependent variable (all variables listed in 2.1). Finally, to estimate the strength of the effect of the Swiss Fintech Corner, dummy or intervention variables were added to September 2016, the time the event took place.

# Findings

## Semi-structured interviews

The four people interviewed as part of this case study and the acronyms which will be used to identify their responses throughout these findings *[xxxx]* are:

- ☐ Anja Loetscher (Director of the Geneva Convention Bureau) – [*Loetscher GCB*]
- ☐ Michael Kleiner (Economic Development Officer, State of Geneva) – [*Kleiner EDO*]

- ☐ Sven Bossu (Head of SIBOS) – [*Bossu SIBOS*]
- ☐ John Hucker (President, Swiss Finance & Technology Association) – [*Hucker SFTA*].

## Question 1

The first question of the interview (see above) addressed the possible effect of the Swiss Fintech Corner on the visibility of Geneva. The lexical analysis shows three slightly different groups.

*Loetscher GCB* and *Bossu SIBOS* show a significant level of homogeneity and a broadly positive evaluation. *Loetscher GCB* has the highest lexical richness in that the terms used referred almost entirely to mass media impact. This shows she was fully aware of the importance of including Geneva in global mass media.

> "For us to have had SIBOS here in Geneva, gave us wide local and international press coverage" (*Loetscher GCB*).

The second group of the lexical component is from *Bossu SIBOS*. For him, the impact of the event on Geneva as a meeting industry destination was not significant.

> "Personally, I don't think the fintech corner had any impact on the visibility of Geneva as a destination and that was not its purpose either" (*Bossu SIBOS*).

Rather, he said, the aim was to showcase Switzerland and Geneva as a fintech hub.

> "The entire idea around the fintech corner was to promote Switzerland as a fintech hub and to connect local fintech companies with the global financial institutions" (*Bossu SIBOS*).

This observation is in line with the response from *Loetscher GCB*.

> "We, as the Convention Bureau, align our work according to the Economic Development Office's clusters – one of them being finance" (*Loetscher GCB*).

The two other responses were similar to each other and form the third group. For *Hucker SFTA* there was an impact, but more on Geneva as a canton than on the city itself.

> "For a destination in general I think people could see just how high quality it was but also in terms of the scale of the city ... it was good to bring the global audience on that scale to a city like Geneva" (*Hucker SFTA*).

Finally, *Kleiner EDO* was more focused on direct impacts, stating the number of overnight stays generated specifically by the Swiss Fintech Corner was limited.

> "The audience we were catering for was very specific; startups - which don't have much to do with the tourism industry in general. They very rarely have large meetings ... they meet in little demo days and pitching days, but not necessarily large events" (*Kleiner EDO*).

Nevertheless, he does agree the event was important for the startups.

> "For the startups this was quite a unique opportunity to be in such a large event" (*Kleiner EDO*).

Essentially, interviewees felt the Swiss Fintech Corner's impact on Geneva's image as a meeting industry destination was limited but that there was an impact on Switzerland as a fintech hub connecting "local fintech companies with the global financial institutions" (*Bossu SIBOS*).

## Question 2

The second question aimed to study the inverse effect; how Geneva, as a global financial market, affects the visibility and/or synergies of startups. *Loetscher GCB* did not answer this question, *Bossu SIBOS* gave a limited answer of only 18 main lexical elements, whereas *Kleiner EDO*'s number was 70 and the *Hucker SFTA*'s was 145. The big imbalance in the number of lexical elements invalidated the statistical analysis of homogeneity.

> "The fact that Geneva is an important banking marketplace was one of the reasons why SIBOS came to Geneva" (*Bossu SIBOS*).

Regarding the increasing awareness among the public and political authorities of the existence of fintech startups and their importance, *Bossu SIBOS* thinks it was positive.

> "In that sense, it did boost awareness with the public opinion and local political authorities" (*Bossu SIBOS*).

In line with this opinion, *Hucker SFTA* sees an impact on public opinion and recognises the role of Geneva in this impact.

> "I think Geneva and fintech are sometimes surprised to hear there's a startup community. It's something that could be further developed and improved ... I think the fact it's a financial centre is a huge strength" (*Hucker SFTA*).

*Kleiner EDO* has a slightly different opinion.

> "Public opinion hasn't changed much, but awareness of the fintech startups among traditional Genevan financial actors has increased. The link between fintech and SIBOS was made but I don't think this really made a big difference in terms of public opinion. What was interesting was increased business awareness because our main issue was to sensitise the bankers, the financial actors, about fintech ... We invited Swiss bankers; private bankers from Geneva for a fintech corner event" (*Kleiner EDO*).

He said this created links between bankers and the startups, which was the political aim.

Switzerland has two financial centres, Zurich and Geneva. Their repositioning was generally rejected but there was consensus regarding the positioning of the whole of Switzerland at the international level.

There was broad consensus Geneva provided good opportunities to link startups to traditional actors.

> "[Switzerland] is a huge financial centre and there is demand for innovation. Most banks are not necessarily innovative, they were traditional in a lot of ways and the cultures weren't about rocking the boat or doing things differently. I think that's why not only is it about startups doing things but it's about how startups enable corporates to do things ... where you marry tradition with innovation" (*Hucker SFTA*).

## Question 3

The third question examined the effect of the Swiss Fintech Corner on the Geneva start-ups' ecosystem. The question wasn't answered by *Loetscher GCB* but the statistical analysis of the other three showed heterogeneity of the lexical elements. The positive role Geneva has as an ecosystem yielded different strengths of consensus. *Kleiner EDO* partially credits the Swiss Fintech Corner but focuses more on the SIBOS conference overall than on the Swiss Fintech Corner event.

> "We weren't the only startup corner. There were representatives from Belgium, Luxemburg and Singapore. So people (had) opportunities to walk around and meet other startups" (*Kleiner EDO)*.

*Bossu SIBOS* and *Hucker SFTA* had the most positive opinion about Geneva as a fintech ecosystem.

> "The global financial community loved the idea because it allowed them to connect directly to smaller companies with bright ideas on big issues. The feedback I got from the Swiss fintechs was really positive as well. They were given access to the global banks, privileged contacts with very senior representatives, and networking events and presentations. The startup corner was something we'd never done before and it proved very successful ... As for whether it creates competition or synergies: we actually saw both" (*Bossu SIBOS)*.

For *Hucker SFTA*, the focus it placed on Switzerland as an innovative fintech ecosystem was important.

> "Pascal Bouvier, a prolific social media influencer, said he was going to Switzerland to find things out, in part because he'd heard stuff through his Innotribe network, because he was at SIBOS every year and because people told him about some of these companies in the Crypto Valley or in the blockchain community. He didn't actually invest, but he was involved in a syndicate that invested in Knip which was a well-known Zurich startup ... There's events, an association and things like that but a big international event like SIBOS and a platform like the Swiss Fintech Corner in particular gets the best 24 startups all in one place and they get to see each other in a really high level environment. Also, what's great is they get to not only represent themselves and 'wrestle shoulders' against each other but they stand 'shoulder-to-shoulder' representing Switzerland which was really great" (*Hucker SFTA*).

## Question 4

All the interviewees other than *Bossu SIBOS* responded to this question, which is why the heterogeneity analysis will not be included in the analysis. *Loetscher GCB* highlighted the innovative character of the Swiss Fintech Corner at the SIBOS conference and also linked the success of the events to Geneva's importance as a financial marketplace.

> "Organisers will prefer to come to the congress if the topic is relevant to the destination" (*Loetscher GCB*).

*Kleiner EDO* noted the good coordination efforts of the cantonal authorities, with *Loetscher GCB* having brought the event to Geneva. *Hucker SFTA* noted the coordination between different stakeholders (the "special extraordinary effort") was key to success.

> "I think the Swiss Fintech Corner was a fantastic example of the power of a coordinated effort to really highlight the economic benefit and the commercial opportunities for Swiss stakeholders in an international event" (*Hucker SFTA*).

The final question explored ideas for improvement. *Loetscher GCB* and *Kleiner EDO* identified some challenges with local tourism actors. *Loetscher GCB* felt one problem was cost.

> "Our pricing was on average too high" (*Loetscher GCB*).

While *Kleiner EDO* acknowledged there could be improvements around logistical issues.

> "I mean there was a post-mortem, you know, discussion with the hotels, with the public transport, with all the different actors" (*Kleiner EDO*).

The Economic Development Office did try to reduce the participation cost for startups.

> "In the end, thanks to the sponsors we found, we could reimburse the startups with the amount they had given us so it cost them hardly anything" (*Kleiner EDO*).

*Bossu SIBOS* wants to improve governance by expanding their data collection before and during the event to better assess impacts.

> "To be changed: make sure governance is clear from start and set up the follow-up study before starting the event, allowing us to collect data during the event" (*Bossu SIBOS*).

## Swiss Fintech Corner Startups survey

Of the 24 startups involved in the corner, only 16 responded to the satisfaction survey, with some providing incomplete responses. Just under 50% said they'd spent two days at the event, while the majority spent only one day there. The mean overall satisfaction level was 4.43 on a rating scale from 1 (unsatisfied) to 5 (extremely satisfied), so most were either very satisfied or extremely satisfied.

- Some points which participants felt could be improved included:
- Earlier preparations;
- More media coverage and focused articles on the startups;
- Having private banks visit each start-up booth;
- Giving startups opportunities to participate in innovation talks;
- Better location of the corner and more space for each start-up.

Despite these suggestions, the following performance indicators each received an average score above 4:

- Preparation of the associations involved in organising the Swiss Fintech Corner;
- Communication;
- Location of event;

- ☐ Design and layout;
- ☐ relevance of participation.

## Social media metrics analysis

The hashtag #swissfintechcorner was only used during the event. Therefore, some conclusions below should be considered with caution.

The Twitter analysis of the Swiss Fintech Corner effects yielded the following conclusions.

First, no positive effect was observed on new followers and average engagement rate. In the former case, the number of followers steadily and exponentially increased. In the latter case, unfortunately, the trend showed a slow decrease, with a negative effect immediately after the event (October 2016).

Second, tweets, re-tweets and link clicks showed a positive punctual effect. Punctual means the effect only persisted during September 2016, then disappeared. Tweets and re-tweets added effects on average were 65% and the development of link clicks in September 2016 was half what would have been observed if the Swiss Fintech Corner had not occurred.

Third, for both total tweet impressions and likes, the effect was positive for September 2016. In both cases, the Swiss Fintech Corner doubled the long-run trend. Unfortunately, the positive effect was short-lived, with the numbers decreasing during the four months following the event.

# Discussion of findings

## Findings from interviews

One of the main difficulties in analysing these findings was the question of how to separate impacts perceived by the interviewees to have occurred because of the Swiss Fintech Corner (the focus of this research) and impacts perceived to have been achieved because of the larger SIBOS conference itself. Many responses provided indicated the larger conference was clearly linked to many benefits, but isolating benefits derived from the Swiss Fintech Corner was more difficult for interviewees. In addition, the interviewees were shown to have notably different points of view based on their position in relation to the business event and their overall profession. Nevertheless, benefits related to the Swiss Fintech Corner were concluded based on three levels, which follow below.

### *On Geneva as a meeting industry destination*

While demonstrating impacts on Geneva as a meeting industry destination via the Swiss Fintech Corner is challenging, the SIBOS conference has been seen to further the image of Geneva as an important financial centre. The fintech hub relevance was not limited to Geneva but instead spread over Switzerland.

In terms of interviewees' perspectives on the impact of the Swiss Fintech Corner on Geneva as a destination, responses were varied. The cluster policy linking local economic development and knowledge (research and development) and the meeting industry is fruitful for a finance innovation sector such as fintech.

The policy of the Geneva Convention Bureau is predominantly to market Geneva as a meeting industry destination in relation to the identified local clusters of excellence. In this way, the benefits dynamic is a feedback loop. The local industry cluster of Geneva spreads its excellence (knowledge, research and development, innovation, etc.) to the international audience participating in the business event, while the industry activities of the business event provide benefits to Geneva's knowledge and research and development, and add valued skills. This policy generates a dynamic feedback loop between Geneva's economic and knowledge environment and the industrial sector involved in an event. In other words, the meeting industry policy is a tool for promoting Geneva, not only as a destination but also as a center of economic, knowledge and innovation excellence.

### *On the macroeconomic level*

There were contrasting points of view on this topic, the most contrasting being that of *Kleiner EDO*, whose responses focused almost entirely on the main business event (SIBOS) rather than the Swiss Fintech Corner. Another demarcated opinion was that of *Loetscher GCB*, who was more focused on the development of Geneva as a successful meeting industry destination. For *Loetscher GCB*, the importance of global mass media coverage was significantly higher, which is understandable given its links to the image of a destination.

According to *Hucker SFTA*, traditional bankers were able to connect with and be made aware of the existence and importance of fintech for the financial sector. The Swiss Fintech Corner also enabled the creation of links between the local start-up industry and the local finance industry (*Kleiner EDO*).

In relation to Question 3, the government representative indicated greater interest in the 'big picture', namely the SIBOS conference overall, whereas for *Bossu SIBOS* and *Hucker SFTA* the focus was on the positioning of Switzerland (not specifically Geneva) as a fintech ecosystem. For them, the business event was a real success in this regard.

### *On the microeconomic level*

The startups involved in the Swiss Fintech Corner generally made positive comments about the event, with overall satisfaction ranging from good to excellent. Suggested improvements included:

- Better advance organisation;
- Better location within the conference layout;
- More advertising and media coverage;
- Inclusion of other pull factors to attract the right visitors;

- ☐ A more egalitarian participation of the startups in regard to participation in interviews and innovation talks.

Interviews and survey analysis documented stakeholders' optimism that the Swiss Fintech Corner will contribute to the future success of the start-up companies through the event-related exposure/visibility and direct contact with audiences, which may lead to partnerships/collaborations. Despite the fact that some startups showed a highly successful outcome in terms of social media coverage, converted leads, and exposure in general, these benefits were not distributed evenly among the startups surveyed as some startups were not directly involved in the banking sector.

The hashtag created and used during the business event (#swissfintechcorner) was able to drive meaningful traffic to The Swiss Finance & Technology Association in terms of social media exposure for the association, specifically with regard to Twitter.

Regarding the amount of media coverage received by the startups, the main platform was social media, representing over 60% of the total media coverage. Seventy-five pieces of media coverage related to 10 of the participants; however, just under half of this social media coverage was linked to a single participant. The online press represented 22% (27 pieces) of total media coverage, with nearly half attributed to that same single participant. The print press, radio and TV followed with considerably lower figures.

It must be noted that this question can be interpreted in many ways, especially with regard to social media where the same post is shared multiple times and the inclusion or exclusion of re-tweets may significantly affect the figures provided. Therefore, there is not enough data to interpret the effectiveness of each type of media coverage.

Even though the startups expressed high satisfaction regarding social media coverage, the positive effects were difficult to maintain after the hashtag was no longer used. The lesson is that the event does increase visibility (retweets, likes, link clicks), but making these effects last over time is still a challenge. One year or even less is not enough to measure or filter seasonal effects or patterns on these results. The analysis of social media metrics during other business events will be necessary, not only to confirm this fact, but also to find strategies to cope with this issue.

Only nine technical leads were generated. A possible reason for this may be that 'getting technical leads' was not the aim of all the startups present and the audience may not have been suitable. However, participants had, on average, five commercial leads each. The quality of commercial leads was rated as average with a score of 3.44 out of 5. A possible hypothesis for this rating may be that such commercial leads are considered as early leads and not converted leads. Missing information, such as the specific field of each fintech participant and the age of the startup company, may explain some of the above-mentioned factors.

## Conclusions

There does appear to be a general consensus in the responses provided during this research: that hosting SIBOS in 2016 led to positive legacies in terms of improving Geneva's profile as a meeting industry destination and Switzerland's image as an innovative and leading centre of financial technology excellence. However, isolating and identifying the positive legacy outcomes of the Swiss Fintech Corner, itself, were more difficult. This may have been because the tools and the timing of the research needed improvement.

The findings of this study did not yield enough evidence to support the assumption that "Organisers will prefer to come to the congress if the topic is relevant to the destination". It appears that it may depend on the kind of event, such as the field (i.e. scientific, industrial), or the customer (i.e. corporation, association). Furthermore, even if this is correct, it cannot supersede other factors such as competitive position in price/value, access, additional services, leisure facilities (Boo *et al.*, 2008) or the quality of supplier management (Weber & Ladkin, 2003). In fact, the latter statement about the relevance of the business event topic to the destination appears as a push factor; however, the authors believe that it was a pull factor. Some interviewees pointed to the difficulty in engaging with different providers, such as the accommodation or transportation sector, despite the benefits this business event yielded.

Finally, evidence suggests a dynamic feedback loop was existent. As noted in the benefits section, some startups were seen to gain opportunities, while the Swiss Finance & Technology Association was able to increase its visibility. Moreover, while Geneva may not have gained visibility specifically, the profile of the fintech ecosystem of Switzerland as a whole was raised.

## Limitations of the research

The primary data, specifically the interviews conducted, had a number of limiting issues. Some questions were not answered by all interviewees, possibly because of difficulties in interpreting what information was required. The questions were all very open-ended, making establishment of consistencies in responses challenging. There was also the issue of one of the recorded interviews failing, and only written responses being available.

There were also issues in terms of the quantitative data. Firstly, only 16 of the 24 fintech startups responded and there was no follow-up to gather the remaining responses. Further, the significant time period between the event and the survey meant respondents had to rely on recall, which could have changed the evaluation. The time series evolution metrics used in the social media metrics analysis was less than one year; therefore, some sophisticated statistical tools, namely the structural time series, could not be applied and seasonality patterns could not be filtered.

## Future research

Further research into potential legacies arising from initiatives such as the Swiss Fintech Corner is warranted. Such research, however, would benefit from being incorporated into the planning stage. A Delphi survey could be developed to measure the strength of the perceived benefits at the microeconomic level, with all the fintech startups being involved. This would enhance the potential of evaluating the actual Swiss Fintech Corner outcomes far more effectively than this study is able to do.

From an academic point of view, it would also be useful to improve the quality of the follow-up with the startups to see the evolution, if any, that may be related to the impact of such an event.

# Theoretical discussion

Beyond identifying the general and long-term legacies of the business events industry in a particular destination, it is also vital to gain a better understanding of the impacts such events might have upon *a particular industry sector* within that destination. Of special interest is the extent to which a particular business event may contribute to knowledge creation/innovation networks within particular industries, both within and between organisations. It is also useful to consider how business events more generally can drive innovation and entrepreneurialism.

Governments around the world have devised national innovation agendas that encourage world-class researchers and businesses to collaborate to shape future industries, drive economic agendas and generate wealth. These are positive initiatives, but how might business events contribute to their achievement? A small number of studies have established links between play and business events (Foley *et al.*, 2014; Jonson *et al.*, 2015), noting how integrating concepts of playfulness into business event environments might facilitate entrepreneurial thinking or innovation. Crowther *et al* (2018) provides a useful examination of the role of play in creation of entrepreneurial outcomes during business events, the characteristics of a playful event environment, and how the physical space influences playfulness in an event context. They use the example of a Danish lighting festival to demonstrate how business events can offer the space, time, activities and socialisation opportunities that enable serendipitous discovery, creation and entrepreneurship (via moments of insight sparked as a result of a creative process) (Crowther *et al.*, 2018).



Business events offer an "ideal platform for this collaboration providing transformative social networks and attracting global talent to further enhance the co-creation of value for the community at large" (Edwards *et al.*, 2016: 6). Innovation is increasingly dependent on collaboration, which relies on the camaraderie, trust and relationships formed through face-to-face contact. Business events provide the platforms for building relationships and the foundation for the networks that drive collaboration and innovation.

Business events also provide an intentional disruption of time, space and the patterns of ordinary life (Patterson & Getz, 2013). This opportunity to be removed from the mundane, predictable aspects of daily life can create a sense of experimentation, adventure and playfulness in attendees (Foley *et al.*, 2014). Play is a remarkably significant and rich experience which, when combined with enjoyable or 'flow' experiences, can create a highly effective learning environment (Jonson, 1999). Sociologist Georg Simmel's (1964) concept of play and sociability, the "pure, and arguably quintessential, moment when conversation is being conducted for its own sake without any extrinsic motive by either party who are, for that moment, equals" (Jonson, 1999: 18) can be useful in a business event setting (Foley *et al.*, 2014). Essentially, being in a state of play invites exploration and experimentation and can encourage innovation and entrepreneurial thinking during business events (Foley et al., 2014).

The knowledge exchange, problem-solving, focused discussion and opportunities for socialisation and serendipitous meetings or conversations which can occur at well-designed business events, when combined with experiences of play, escape from everyday life, and flow, form the perfect environment for creativity, openness to new ways of thinking and experimentation. Each of these can be considered precursors to entrepreneurial thinking and innovation (Crowther *et al.*, 2018), so it is important that business event designers develop the skills needed to create and recognise such opportunities and experiences in their event design.

The link between innovation, global knowledge networks and business events has been the focus of studies in various distinct yet related fields. Richards (2015) argued that in our increasingly interconnected world, business events begin to take on new roles in connecting people, cultures and ideas, while also connecting local and global realms. He highlights recent shifts in social theory which suggest, rather than simply being objects to be studied, business events can also be seen as subjects, or actors, in their own right, capable of playing a role in social systems. Thus, research adopting social-network analysis has risen in prominence in examining business events, given their ability to embed social networks and link key actors together, by mapping nodes supported by business events, as well as the ties business events can facilitate between social actors (e.g. Jarman *et al.*, 2014). Yet, the long-term legacies of networks established at business events are rarely tracked, evaluated or examined.

More research is required to examine longer term effects of business events – not just in terms of individual events, or even a destination's business event portfolio as a whole (e.g. Ziakas, 2014) – but also in terms of the local-global networks that shape and are shaped by those events. To that extent, Podesta and Richards (2018) examined the knowledge spillovers to a region resulting from a cultural festival in Italy, highlighting the potential of this for both destination development and local innovation. Business events can stimulate circulation of knowledge and ideas by creating new relationships and exchanges that can impact both individual and collective learning (De Geus *et al.*, 2016).

Repeated staging of such knowledge-based business events can also embed the knowledge in the host city or region, increasing local distinctiveness and innovation. Podesta and Richards (2018) drew on earlier work by Castells (2009, 2010) who argued that local places gain power by being well connected to the global *space of flows*, and that the global flows of culture are in turn transformed by their concrete use in the *space of places*. Castells also notes that key actors, or *switchers*, who are able to link different global and local networks, play a critical mediation role. Switchers, who are central to the creation, development and eventual transformation of business events, were also identified by Hardbord (2002) when analysing the Berlin Film Festival, and Rovira (2010) when examining Barcelona's design field.

The notion of local-global networks elevating a destination as a hub in the wider knowledge economy has also been assessed in the context of creative festivals. Drawing on the Sonar Festival that originated in Barcelona, Colombo and Richards (2017) demonstrated how a single business event developed in one city can be a source of global, as well as local, impact; shaping not only the city in which it originated, but also stimulating innovation at a global level. In the current competitive global economic environment, it is critical for cities to develop a *hub* role in a knowledge network. In contrast, other locations simply act as *nodes* in global networks that receive knowledge created elsewhere.

Startups are an integral part of major cities' knowledge networks and a vital engine for their development, partially determining whether they will attain a *hub* or *node* status. They help drive innovation, encourage competition, and create employment opportunities (e.g. Koster & Stel, 2014). According to the Center of American Entrepreneurship (2018), following the global financial crisis in 2009 that brought a new period of technological innovation, global venture capital investment surged from $52 billion in 2010 to $171 billion in 2017. During the same period, venture deals grew from around 8600 in 2010 to more than 14,800 in 2017; an increase of nearly 75%. Concurrent with the increase in global venture capital investment was a substantial globalisation of start-up activity and venture capital, with the once dominant United States being increasingly challenged by the rapid ascent of start-up cities in Europe, China and India (Florida & Hathaway, 2018).

The new and innovative ideas behind many startups can also play a role in moving the public sector forward, by providing huge efficiency gains in public services and by helping establish competitive regulatory environments. Further, it is wise for governments to invest in protecting the knowledge economy ecosystems they helped create (Startup Genome, 2020). Preserving the number of startups in an ecosystem helps protect overall economic value because many startups offer higher job multiplier impacts via higher wages and exports and attraction of foreign investment (Start-up Genome, 2020). However, a 2020 global survey of start-up founders indicates that startups need support to attract investment.

Indeed, startups face a variety of challenges, including limited resources, minimal access to experienced mentors, reduced capacity to hire a suitable team and difficulties in attracting buyers without a proven reputation or customer base (Riani, 2020). During times of economic crisis, such challenges become more pressing. According to Statista (2021), in 2006 there were just over 700,000 entrepreneurs in the United States. The number dropped significantly during the Global Financial Crisis, and it took a decade for the number to climb back up to 740,000 (Statista, 2021). According to Start-up Genome (2020), whilst many big-name startups successfully pivoted during COVID-19 because of the resources they had access to, smaller startups struggled to survive because of a 20% decrease in global venture capital funding. Giones *et al.* (2020) note that the intensity of Europe's economic activity freeze meant corporates shifted their focus to internal operational challenges and away from future innovation projects. Further, the focus of many public support programs shifted away from startups as areas such as healthcare, emergency services, and the protection of established organisations were prioritised (Giones *et al.*, 2020).

Essentially, disruptions (such as the Global Financial Crisis and COVID 19) mean viable entrepreneurial startups lose opportunities for visibility, support and financial investment. This includes the emotional support so often needed by startups as a negativity buffer. Such emotional support is traditionally exchanged via in-person meetings, events, interactions and networks, ecosystem cohesion and a collective culture, but with fewer face-to-face opportunities, as well as fewer policies targeting new ventures, there is concern that entrepreneurs are experiencing increasing demand for, but decreasing supply of, emotional support in the way that it is traditionally exchanged when they need it most. Thus, the role of business events in providing startups with platforms for visibility and networking with potential customers and investors (as shown in the case study) becomes even more valuable in times of crisis.

## Lessons learned

This chapter has presented a case study and theoretical viewpoints that highlight the need for a shift in perspective on business events. This shift demands examination of longer term impacts for both individual business events and a destination's business event portfolio, particularly in terms of the knowledge/innovation networks that can shape and are shaped by those business events. The chapter underscores the value of destinations in strategically bidding for business events which link local-global knowledge networks (including startups) as a catalyst for innovation and development to occur.

To address the challenges noted in the previous section, extensive networking is key to the survival of startups, which means business events (in the form of conferences, exhibitions or tradeshows) can be critical in connecting emerging entrepreneurial talent and ideas with investors, suppliers and clients, as was the case with the SIBOS conference.

Existing research indicates that industry ecosystems that include diverse technologies and channels through which knowledge is effectively exchanged support the creation of startups. Those with processes through which both formal and informal relationships can be formed also benefit startups, particularly those with processes in place to encourage the formation of relationships between startups and external partners, especially when those external partners are in the public sphere (Spender *et al.*, 2017). Well-planned and strategically designed business events and initiatives like the Fintech Startup Corner are particularly apt examples of such processes, suggesting potential for business events to be integral to network creation for startups, and hence, for driving innovation.

It should also be noted that such benefits are amplified when business events are held in destinations which are actively investing in the creation of extensive and innovative eco-systems. These eco-systems can attract global talent with a forward-looking mindset, and even serendipitous encounters at a coffee shop can result in advancing prospects for new ideas, startups, or a field as a whole (Feld & Hathaway, 2020).

Startups are integral to knowledge networks and key drivers of innovation. As demonstrated in the case study, business events serve as ideal social spaces to link startups with the more established players in an industry sector, bringing benefits to multiple stakeholders.

Finally, we looked at the potential for business events to generate and support innovation, and creative ecosystems more generally. Sociability, trust and creation of a shared social reality during business events underpin potential for entrepreneurial outcomes to be achieved (Foley *et al.*, 2014). This is because individuals are able to detach from their own personal and social constraints, to find themselves in an artificial environment of temporary equality which gives them the freedom to experiment, take risks and engage in creativity. Crowther *et al* (2018) state purposeful event design should include times and spaces within the event schedule that enable participants to establish shared meanings; be spontaneous; enjoy humour, impulsiveness and unpredictability; socialise; absorb and generate 'energy' via the buzz of the event, and experience flow. They argue that such experiences underpin the interrelationship between play and events, and are crucial to unlocking the potential which business events hold to prompt an entrepreneurial mindset in attendees.

Essentially, business event designers who are able to use play, sociability, intimacy and flow to create environments which not only facilitate knowledge sharing, but also include safe opportunities for semi-guided experimentation, risk-taking and collaboration will encourage innovative ideas and an entrepreneurial mindset in attendees. This then has the potential to generate opportunities for successful partnerships (startups) between innovators and those seeking formation of innovative ideas. If the achievements and benefits generated by entrepreneurial startups initiated at particular business events are then tracked and measured over time, the value proposition of the events industry will be expanded and more accurately understood.

## References


BAK Basel Economics AG (2015) *The economic significance of the Swiss Financial Sector*, Report prepared for the Swiss Bankers Association and the Swiss Insurance Association.

Boo, S., Koh, Y. and Jones, D. (2008) An exploration of attractiveness of convention cities based on visit behavior, *Journal of Convention & Event Tourism*, **9**, 239-257.

Castells, M. (2009) *Communication Power*, Oxford: Oxford University Press.

Castells, M. (2010) *The Rise of the Network Society. The Information Age: Economy, Society and Culture: Volume I*, Malden, MA: Wiley-Blackwell.

Center for American Entrepreneurship (2018) America's rising startup communities. Retrieved 10 May 2021. https://startupsusa.org/

Colombo, A and Richards, G. (2017) Eventful cities as global innovation catalysts: The Sónar festival network, *Event Management*, **21**, 621-634.

Crowther, P., Orefice, C. and Beard, C. (2018) At work and play: Business events as entrepreneurial spaces, *The International Journal of Entrepreneurship and Innovation*, **19** (2) 90-99.

De Geus, S., Richards, G. and Toepoel, V. (2016) Conceptualisation and operationalisation of event and festival experiences: Creation of an event experience scale, *Scandinavian Journal of Hospitality and Tourism*, **16**, 274-296.

Deloitte (2016) *Connecting Global Fintech: Hub review 2016*, report prepared for Global Fintech Federation.

Edwards, D., Foley, C. and Hergesell, A. (2016) *Conferences: catalysts for thriving economies*, report prepared for Business Events Sydney, University of Technology Sydney. https://www.besydney.com.au/media/1611/catalysts-for-thriving-economies-research.pdf

Edwards, D., Foley, C. and Malone, C. (2017) *The Power of Conferences: Stories of Serendipity, Innovation and Driving Social Change*, Sydney: UTS ePress.

Ernst & Young (2016) *Swiss FinTech report 2016: the role of Switzerland as a financial hub*, report prepared for Swiss Finance and Technology Association.

Feld, B. and Hathaway, I. (2020) *The Startup Community Way: Evolving an Entrepreneurial Ecosystem*, New Jersey: John Wiley & Sons.

Florida, R. and Hathaway, I. (2018) How the geography of startups and innovation is changing, *Harvard Business Review*. https://hbr.org/2018/11/how-the-geography-of-startups-and-innovation-is-changing

Foley, C., Edwards, D. and Schlenker, K. (2014) Business events and friendship: leveraging the sociable legacies, *Event Management*, **18** (1), 53-64.

Foley, C., Schlenker, K., Edwards, D. and Hayllar, B. (2010) *A scoping study of business events: Beyond tourism benefits*, report prepared for Business Events Sydney, UTS, Sydney. https://opus.lib.uts.edu.au/handle/10453/16848

Foley, C., Schlenker, K., Edwards, D. and Lewis-Smith, L. (2013) Determining business event legacies beyond the tourism spend: An Australian case study approach, *Event Management: An International Journal*, **17** (3), 311-322.

Giones, F., Brem, A., Pollack, J., Michaelis, T., Klvver, K. and Brinckmann, J. (2020) Revising entrepreneurial action in response to exogenous shocks: Considering the COVID-19 pandemic, *Journal of Business Venturing Insights*, **14**, e00186.

Hansen, N. (2010) *Conferences as dramaturgical learning spaces*, Doctoral Thesis, Danish School of Education, Aarhus University.

Hardbord, J. (2002) *Film Cultures*, London, UK: Sage.

Jago, L. and Deery, M. (2012) *A theoretical framework for business events: The role of business events in innovation and knowledge creation*, CAUTHE conference proceedings (2012), Victoria: La Trobe University.

Jarman, D., Theodoraki, E., Hall, H. and Ali-Knight, J. (2014) Social network analysis and festival cities: an exploration of concepts, literature and methods, *International Journal of Event and Festival Management*, **5** (3), 311-322.

Jonson, P. (1999) Play and 'enjoyable experience': A pedagogical framework, *Moritz*, **5** (1), 18-21.

Jonson P., Small J., Foley C. and Edwards, D. (2015) 'All shook up' at the Parkes Elvis Festival: the role of play in events, *Event Management*, **19** (4), 479-493.

Koopman, S, Harvey, A, Doornik, J and Shepard, N. (2009) *STAMP 8.2: Structural Time Series Analyser, Modeller and Predictor*, London: Timberlake Consultants.

Koster, S. and Stel, A. (2014) The relationship between start-ups, market mobility and employment growth: An empirical analysis for Dutch regions, *Papers in Regional Science*, **93** (1), 203-217.

Patterson, I. and Getz, D. (2013) At the nexus of leisure and event studies, *Event Management*, **17** (3), 227-240.

Podestà, M. and Richards, G. (2018) Creating knowledge spillovers through knowledge-based festivals: the case of Mantua, Italy , *Journal of Policy Research in Tourism, Leisure and Events*, **10** (1), 1-16.

Riani, A. (2020) 5 Big challenges to launching a startup and how to overcome them, Forbes. https://www.forbes.com/sites/abdoriani/2020/01/30/5-big-challenges-to-launching-a-startup-and-how-to-overcome-them/?sh=1b07636f3364

Richards, G. (2015) Events in the network society: The role of pulsar and iterative events, *Events Management*, **19** (4), 553-566.

Rovira, E. (2010) *BCN how are you?*, MA Thesis, Tilburg University, Tilburg, Netherlands.

Scaglione, M. and Dimitriou, S. (2018) Zoom-in on the first-ever Swiss Fintech Corner. University of Applied Sciences and Arts Western Switzerland Valais. https://www.themeetingsindustry.org/web/wp-content/uploads/JMIC_Case-Study-Program_Geneva_Swiss-fintech-at-Sibos-17.5.pdf

SIBOS (2017) Analysing the Sibos Experience, retrieved 7 March 2017 from https://www.sibos.com/conference/hub/articles/analysing-sibos-experience

Simmel, G. (1964). *The Sociology of Georg Simmel*: Translated, Edited. Free Press of Glencoe.

Spender, J., Corvello, V., Grimaldi, M. and Rippa, P. (2017) Startups and open innovation: A review of the literature, *European Journal of Innovation Management*, **20** (1), 4-30.

Startup Genome (2020) The global startup ecosystem report 2020: The new normal for the global startup economy and the impact of COVID-19, GSER, https://startupgenome.com/article/save-your-startups-using-policy-to-kick-start-post-crisis-economy

Startupticker.ch. (2016) Making Geneva a global fintech hub. https://www.startupticker.ch/en/news/july-2016/making-geneva-a-global-fintech-hub

Statista (2021) Number of business establishments less than 1 year old in the United States, March 1994 to March 2020. https://www.statista.com/statistics/235494/new-entrepreneurial-businesses-in-the-us/

Swiss Finance and Technology Association. (2016), Sibos comes to Geneva – Exhibit at Sibos! retrieved 7 March 2017 from https://swissfinte.ch/sibos2016/ .

The Iceberg (2017) Geneva: Sibos 2016 and aligning with knowledge sectors. https://www.the-iceberg.org/features/congress-legacies-association-mission-destination-vision-legacy/.

Veal, A., Darcy, S. and Lynch, R. (2012) *Australian Leisure*, 4th edn, Melbourne: Pearson Higher Education.

Vlasov, S., Bahlmann, M. and Knoben, J. (2017), A study of how diversity in conference participation relates to SMEs' innovative performance, *Journal of Economic Geography*, **17** (1), 191-216.

Weber, K. and Ladkin, A. (2003) The convention industry in Australia and the United Kingdom: Key issues and competitive forces, *Journal of Travel Research*, **42**, 125-132.

Ye, P., Liu, L. and Tan, J. (2021) Influence of knowledge sharing, innovation passion and absorptive capacity on innovation behaviour in China, *Journal of Organizational Change Management*, https://doi.org/10.1108/JOCM-08-2020-0237

Z/Yen Group (2016) Global Financial Centres Index, retrieved 10 March 2017 from https://www.longfinance.net/programmes/financial-centre-futures/global-financial-centres-index/

Ziakas, V. (2014) Planning and leveraging event portfolios: Towards a holistic theory, *Journal of Hospitality Marketing & Management*, **23** (3), 327-356.

# 9 Business Events and Diplomacy

> *Conferences are the soft power assets that allow us to sell Sydney and NSW to the world and, following the global pandemic, this is now more important than ever.*
> Lyn Lewis-Smith, CEO, Business Events Sydney

From a narrow perspective, business events may be seen as isolated events occurring within particular industry sectors (i.e. science, business, agriculture, education, finance etc.). However, there is increasing awareness that, in reality, they are multifaceted assets to economies, government entities, communities and industries, and that they are able to make significant contributions to broader societal and diplomatic outcomes. Business events not only deliver direct economic benefits (via delegates occupying hotel rooms and spending tourist dollars) (Foley *et al.*, 2019) but serve to strengthen international relationships in respect of both delegates and institutions, showcase local expertise (Foley *et al.*, 2010; Foley *et al.*, 2013), attract investment and trade, bring focus to global issues (Edwards *et al.*, 2011), advance international research agendas (Edwards *et al.*, 2017), and promote places and locations to strengthen place branding and attraction (Edwards *et al.*, 2016).

The following case study considers the role of a major international business event, the 11th China-LAC Business Summit, in creating and/or establishing mutually beneficial trade partnerships among government representatives, investors and business operators from China and Latin American (LA) and Caribbean (C) countries. It also discusses the contribution the event made toward raising the profile of the host city, Punta del Este, as a meetings industry destination and the ways in which it acted as a catalyst for regional economic growth and development in Uruguay.

Whilst there is considerable anecdotal evidence suggesting the event was impactful in each of these areas of legacy, capturing and measuring the extent to which it contributed is more difficult. There were no processes, tools or human resources allocated toward formally capturing research data in the planning phase or from participants during the event itself. It was only decided nearly two months after the event had concluded that it would be useful to gather some primary data (interviews) and use existing secondary data (reports and statistics) to collate a report. Even then, few resources and a very short timeframe were allotted, which meant the primary data were limited to interviews with event organisers.

Despite these challenges, the information upon which the case study is built still offers valuable insights into the ways in which the China-LAC Business Summit contributed to legacy creation for those parties involved. It also offers a useful base on which similar studies may build. Future researchers will benefit from the perspectives it presents in terms of how the business event was positioned and executed to achieve specific legacy outcomes. Industry practitioners and event professionals will also benefit from the opportunities it offers to deepen their understanding of diverse markets and subsequently create similar outcomes in other settings. Against the background of this case study, soft power, and particularly the links between a country's foreign policy, culture and business events, is discussed in the theoretical considerations section, with specific reference to China.

## Case study: Investing in business events for regional growth and development

| | |
|---|---|
| **Business event** | 11th China-LAC Business Summit 2017 |
| **Location** | Punta del Este – Uruguay |
| **Type of event** | Commercial/trade business summit |
| **Business event purpose** | To improve trade and investment pathways and opportunities between the countries involved |
| **Primary audience** | Chinese, Latin American and Caribbean developers, business representatives, government (local, regional and national), investors, planners and the media (2315 delegates in total) |
| **Primary business event legacy** | Generated and consolidated strategic commercial ties between investors, government organisations and businesses from China, Latin America and the Caribbean to support economic development and infrastructure growth |
| **Additional legacies (beyond tourism)** | Promotion of Punta del Este and surrounding regions as a business meetings and tourism destination; creation of a more stable local economy as a result of 'de-seasonalising' the tourism industry; more effective allocation of available global resources in major infrastructure projects |
| **Summary quote** | "(This business event) was and still is a powerful tool, showcasing Punta del Este at an international level. It's a perfect example of the benefits that countries and communities can obtain through international business." |

This case study (CS) was adapted from: Alvez, M. J., Sinigaglia, I., Lourdes, A. and Perrier, G. (2019).

## Background

Punta del Este is a beachside city in southeast Uruguay, Latin America. In summer its resident population of around 10,000 swells to nearly five times that, as thousands of wealthy tourists head to what is sometimes referred to as 'The St. Tropez of South America'. Property prices, rents and the cost of living in the city are incredibly high and there is no public housing. This has forced low and even middle income earners to the fringes, creating a unique and relatively sociologically homogenous community profile and a playground for the rich.

## About the China-LAC business summit

The China-LAC Business Summit is an annual business event organised by the Chinese Council for the Promotion of International Trade (CCPIT), with the cooperation of the Inter-American Development Bank (IADB). The first edition of this highly successful business event took place in Santiago in 2007. It focuses on key issues facing Chinese, Latin American and Caribbean (LAC) business communities in light of broader international economic and trade development trends. Political leaders, experts, representatives of international organisations and financial institutions, as well as entrepreneurs from China and LAC countries, are invited to discuss trade and investment policies, market opportunities, and mutually beneficial cooperation. The summit is held annually in China and LAC countries on a rotating basis and attracts more than 1000 delegates from nearly 30 LAC countries each year.

It is the most important annual business event for businesspeople, entrepreneurs and the governments of China, Latin America and the Caribbean, with regard to the exchange of business and investment prospects. Its 11th edition was held in Uruguay from 30 November to 2 December 2017 and was organised by Uruguay XXI (the investment, export and country brand agency) and Grupo Elis. Uruguay was chosen as host destination of the business event in recognition of the country's long history of trade and diplomatic relations with China. The initiative was identified as an important element of the 2015-2019 China-Latin America Cooperation Plan, promoted by the Community of Latin American and Caribbean States.

The rationale behind establishment of this annual business event was based on regional features that suggested improved relations between the countries would be mutually beneficial.

Features that gave China a place in the relationship were:

- A growth rate significantly higher than the world's average;
- One of the world's greatest food exporters and importers;
- A strong middle class;
- Strategic partner for the promotion of global trade; and
- Great opportunities for complementarity.

Features that gave Latin America and the Caribbean a place in the relationship were:

- One of the most dynamic economies in the world;
- A 600-million strong consumer market;
- An important world reserve of energy and minerals;
- Its reputation as the world's primary food exporter.

According to information provided by the World Economic Forum (Melguizo, 2017), China was already the main business partner of Argentina, Brazil, Chile, Peru and Uruguay, as well as the second most important for Mexico. Another characteristic was China's interest in investing in public development in the region, which is a strategy of the Chinese government.

In an interview published in XinhuaNet in 2019, Veronica Zavala, the Inter-American Development Bank (IDB) representative for Central America, said she believed the summit would serve as an essential tool to generate new business opportunities and promote exports and foreign investment. She said:

> *"Businesses are increasingly global, but they are built on contacts between people at forums (such as the China-LAC Business Summit). We need to open doors to have the opportunities that put us on the path to inclusive growth,"*

Essentially, this entire business event was conceived with two main purposes. First, it aimed to facilitate the generation and consolidation of strategic commercial ties with important investors. Second, it would enable government organisations and Chinese companies to forge mutually beneficial partnerships with entities in Latin America and the Caribbean. In other words, it would provide the participating organisations and governments of these regions with an opportunity to find the support and investors they need for infrastructure and development plans for their destinations.

## Objectives of China-LAC 2017

The program for the 2017 edition of China-LAC was focused on the following issues:

- How is Latin America inserted into the 21st Century Maritime Silk Road?
- What can be learned from the business success experiences thus far between the two regions?
- How do we create more added value and generate more productive linkages?
- What is the role of Chinese currency on international trade, particularly in the LAC region?
- What are the challenges ahead in terms of infrastructure in the new global economy and how will those challenges be addressed in the strategy for the Belt and Road Initiative?
- What kind of opportunities do Chinese banks find in the LAC region?
- What is Latin America's role in terms of food security in a constantly growing China?

- ☐ What cooperation opportunities are available in terms of transport, infrastructure and energy?

In order to adequately address this range of issues, highly reputed keynote speakers and subject matter experts were included in the program, along with influential, powerful and relevant decision-makers from both public and private sectors.

## Attendance figures

Although the initial target for China-LAC 2017 was to attract 500 companies from both regions, the results far exceeded expectations, with 2315 delegates attending. Delegates' countries of origin are shown in Table 9.1.

**Table 9.1:** Number of attendees at the conference by country

| Category/country | Number of attendees |
|---|---|
| China | 614 |
| Uruguay | 824 |
| Lac | 396 |
| Press | 101 |
| Booths | 380 |
| Total | 2315 |

*Source:* Grupo Elis Annex I

There were approximately 1200 scheduled business appointments held between companies and different organisations during the three-day summit. It is at these meetings that individual partnerships and investment deals are arranged.

## Investment opportunities presented during the summit

Countries in Latin America and the Caribbean that were seeking funding from Chinese entities for major developments had the opportunity to present their projects to potential investors. Argentina, Chile and Uruguay each presented details of proposed major development projects for their regions worth hundreds of millions of US dollars. Other LAC countries also brought smaller projects seeking investors and partnerships. These included Ecuador (with a total of USD 14,077.58 million in projects), El Salvador (USD 263.4 million) and Mexico (USD 24,543 million). The complete information of all projects presented for investment was published on the website of the business event.

9

## The Punta del Este Convention & Exhibition Center

The host venue, the Punta del Este Convention & Exhibition Center is the largest public-private investment project undertaken in the tourism sector in Uruguay in the last 50 years (Punta del Este Convention & Exhibition Center, 2017). The original concept for the facility was motivated by a desire to develop Punta del Este and the surrounding region as a leading meetings industry destination, in an effort to 'de-seasonalise' the strength of the local economy. The peak tourism season (December and January) demands ongoing investment in high-quality accommodation and infrastructure, but when tourists leave, the local economy becomes dependent on events (leisure, sporting and business) for stimulation. In order to achieve this, government considered construction of a convention and exhibition centre that met international standards and which had top-level infrastructure suitable for business events, as a priority. The USD 35,000,000 it cost to build the facility appears to be paying off thus far. Since it became operational, the city has enjoyed the highest off-season hotel occupancy rates ever recorded, and hundreds of new jobs have been created in small and medium-sized service businesses. International business events, which previously could not be held in the region, are now common occurrences. The high delegate numbers and media coverage generated by these is raising the profile of the destination for other types of tourism. Importantly, also, they have raised the level of professionalism of local event management and delivery specialists and encouraged growth of new businesses and specialised services that cater to the meetings sector, known to be a highly profitable global tourism market.

The venue includes 5400 sqm of meeting space, 6700 sqm of exhibition space and numerous other rooms and halls. Between July and December 2017, the venue was host to 50 business events and 40,000 visitors, and created 1600 jobs. The China-LAC event occupied the entire venue for its duration, including all meeting rooms and foyers of the convention centre, while the trade exhibition, which comprised 157 booths for 172 participating organisations, filled the exhibition hall.

## Methodology

A combination of sources was used to gather data for this case study, from published and unpublished material and statistics to meetings with key people directly involved in organisation of the business event. More specifically, it included face-to-face interviews, email responses to questions, observations made at the business event by the original case study author, written reports provided by suppliers, and news and media releases.

The information and responses have been used to inform a qualitative account of the business event and the outcomes it was perceived by interviewees to have generated.

## Data sources

Interviews were conducted with organisers two months after the conclusion of the business event. Those interviewed were representatives of three authoring organisations, a representative from the Ministry of Tourism and a representative from the Ministry of Foreign Affairs.

The following section describes the information provided by each key player directly involved in the organisation of China-LAC 2017. It was collected via formal post-event reports, face-to-face meetings and information provided via email.

### Grupo Elis

Grupo Elis provided information on the business event program, the number of registrations/event participants and data on where delegates had come from. They also offered details (places, activities and attendee information) on technical visits that took place at key Uruguayan companies, and on the number of student volunteers used to deliver the business event.

### Uruguay XXI

Uruguay XXI provided facts and figures about the level of press coverage (free publicity) generated by the business event. This included press at a national and international level. They also provided information about how return on investment as a result of media and promotion was calculated, and the approximate monetary value of that promotion.

### Arquitectura Promocional

Arquitectura Promocional organised the trade exhibition element of the summit and provided data on the metres of exhibition space sold, the number of stands booked, the countries from which those exhibitors had come and the people who worked on the stands.

### Interamerican Development Bank

The Interamerican Development Bank forwarded data on the business networking sessions, including the one-on-one meetings between Chinese companies/organisations and the Latin American and Caribbean companies/organisations seeking funding or project resources.

### Punta del Este Convention & Exhibition Center

The Punta del Este Convention & Exhibition Center supplied information on the space used or sold at the business event, other services contracted (such as food and beverage suppliers, audiovisual services, and businesses involved in the venue decoration/branding), as well as news published during and post event. They also supplied excerpts of interviews and video testimonials recorded during the business event by different media sources.

## Press coverage: Measurement of the return on investment

Uruguay XXI and the Uruguayan Government both invested resources to bring this business event to Uruguay. This was done via the construction of an international standard convention and exhibition centre, and also through direct financial and human resource input. The rationale behind this investment was both to generate trade and foreign investment in the region and to raise the international profile of the city and country via media exposure.

It was therefore important to measure the return received on investment (ROI). In order to do this, records were kept of all media coverage of the business event, of Punta del Este and of Uruguay immediately before, during and after the event. Consideration was given to the type of media, the quality and reputation of the news/media source, the circulation (readership/reach) and the length/quality of the coverage. In this way, the ROI of the press coverage could be measured and monetised by considering the average price that would have been paid for similar marketing exposure (i.e. via paid advertising). These price calculations were provided by the Media Center of Grupo M Uruguay, and Argos Monitoring Agency, contracted by Uruguay XXI. This was the base used for the calculation, and according to the same reference sources, it was possible to establish an estimated price for the articles published on international media, which varies from two and a half to four and a half times the price of an article published on national media in Uruguay. To establish a logical framework for calculation, an estimation of the total media received was made. This breakdown indicated approximately 20% of the articles generated would have been otherwise charged at the highest market price, 40% would have attracted a mid-range charge and the remaining 40% would have been charged at the lowest price.

The exchange rate used was that advertised on 18 December 2017 by the Banco de la República Oriental del Uruguay (Uruguay's National Bank). This was $U29.58 Uruguayan Pesos per USD.

## Measurement of the impact of Twitter

The Twitter strategy adopted by Uruguay XXI during China-LAC 2017 was based on three pillars:

- Planning according to the event's program and three main subject areas;
- Planning and working before the event with companies that were taking part; the Interamerican Development Bank (IADB), speakers, and other parties that were capable of producing a multiplier effect with their tweets and posts;
- Designing specific templates according to the different target audiences and objectives previously intended for them.

According to the official Uruguay XXI Twitter account (@UruguayXXI), 130 tweets were released in total. They were released on three separate dates: 30 November (22 tweets), 1 December (96 tweets), and 2 December (12 tweets). These tweets had 190,870 impressions, 408 clicks on their respective links, 585 re-tweets and 927 likes, as well as 16 replies. Table 9.2 shows the distribution of the effect of the tweets on the different days of the business event.

**Table 9.2:** Distribution effect of tweets on different days of the business event

| | 30 Nov | 1 Dec | 2 Dec | Total |
|---|---|---|---|---|
| Tweets | 22 | 96 | 12 | 130 |
| Impressions | 29,422 | 116,323 | 45,125 | 190,870 |
| Clicks on links | 122 | 230 | 56 | 408 |
| Re-tweets | 67 | 384 | 134 | 585 |
| Likes | 146 | 485 | 296 | 927 |
| Replies | 3 | 8 | 5 | 16 |

*Source:* Uruguay XXI Annex III – Twitter)

## Data limitations

One key aspect which significantly impacted the case study findings was the timing of the research (January 2018) in relation to when the business event was held (30 November to 2 December, 2017). The decision to undertake the research was made post the summit and research commenced more than a month after the business event took place. This meant there were limitations in terms of the ability to collect information prior to and during the event. There was no human resource allocation for interviewing key participants. Because of these limitations there is limited primary data.

# Key benefits realised

Benefits of the 11th China-LAC Business Summit were realised in two major areas:

1. For participating countries there was formation of business ties between various enterprises, resulting in potential investment, infrastructure development, new markets and business opportunities.
2. For Punta del Este and Uruguay there was international attention and media coverage promoting the location as a top destination for international business events and broader tourism. There were also more immediate short-term benefits for the local economy as a result of (delegate and organiser) spending associated with this major business event.

To simplify discussion of the benefits, four overarching categories have been identified:

- Business and/or trade-related benefits
- National and international media exposure
- Social media impacts
- Impacts on the local economy.

## Business/trade-related benefits

Main beneficiaries:

- The local and national tourism sectors
- The companies in the private sector willing to do business with china
- The public sector.

The business event attracted a large number of participants and facilitated vibrant interactions likely to lead to relationships and thus future development in both regions.

Key agreements were reached including:

- A memorandum of understanding signed by the Chinese Council for the Promotion of International Trade (CCPIT), the Uruguayan Chamber of Country Produce, and the Uruguayan Union of Exporters. This agreement will be an important instrument for the promotion of Uruguayan products and exports.

9

- Three other agreements with CCPIT were signed during the business event with organisations from Argentina, Cuba and Colombia for trade promotion between them and China (Punta del Este Convention & Exhibition Center, 2017).

After Uruguay was awarded hosting rights for this event, China and Uruguay signed a number of agreements that significantly changed the relationship between them, moving away from just a bilateral commercial relationship and turning to a strategic partnership. Hence, the foundations were laid for a future free trade agreement between both countries (Punta del Este Convention & Exhibition Center, 2017).

Hundreds of business meetings took place throughout the summit, both in the 140 stands of the 40 countries present and in the Business Round organised on the morning of Saturday, 2 December. To close the day, 200 foreign businessmen made technical visits focused on agribusiness to Bodegas Garzón and Olivares, while visits to Zonamérica and Antel Data Center were attended by those interested in learning about Uruguay's corporate, logistical and technical services.

Within the framework of the China-LAC event, a smaller business event was held for 30 trade promotion agencies and trade organisations from 16 China-LAC countries. According to Uruguay XXI 2017, the so-called Trade Promotion Organisations meeting culminated in the signing of the Declaration of Punta del Este, as well as bilateral agreements of cooperation between agencies.

A number of other major initiatives were announced during the business event, including:

- The creation of the Confucius Institute, a joint endeavour between the University of the Oriental Republic of Uruguay and the University of Qingdao in China, to promote learning of the Chinese language and culture in Uruguay.
- The signing of a memorandum of understanding to celebrate 30 years of Sino-Uruguayan commercial relations, authorising Chinese aircraft and airlines to land in Uruguay, either through direct lines or through regional hubs (Fernandez, 2017).
- Lifan Motors' announcement of their intention to expand their services in Uruguay to include the production of electric cars (Fernandez, 2017).
- An IADB USD 100-million loan to the Financial Fund for the Development of the Plata Basin – FONPLATA, which includes Argentina, Bolivia, Brazil, Paraguay and Uruguay. The loan aims to help finance different projects in infrastructure, governance and tourism in the five member countries (Fonplata, 2017).
- An agreement between the Uruguayan National Meat Institute and the Chinese National Certification and Inspection Group on certification of the processes of traceability of the beef exported from Uruguay. In 2017, Uruguay was ranked sixth largest beef exporter in the world after India, Brazil, Australia, the United States and New Zealand. China is now the largest export market for Uruguayan beef (40% of total exports).

The China-LAC Business Summit is a powerful tool, with the 2017 event in Uruguay showcasing Punta del Este at an international level. It is a perfect example of the benefits countries and communities can obtain through international business events, and of the way their people may benefit in a range of areas, from knowledge production to the creation of jobs and cultural exchange, in the medium and long term.

**Photo 9.1:** 11th China-LAC Business Summit 2017: Signing of a memorandum of understanding to celebrate 30 years of Sino-Uruguayan commercial relations

Due to its size and characteristics, the China-LAC Business Summit is a business event that could not have taken place in any other facilities in Uruguay. This demonstrates the importance of having specialised, highly adaptable venues with large capacities and top services and technology in regions aiming to utilise business events as part of their growth and development strategy.

## National and international media exposure

The following information details the impact made by national and international media on the China-LAC Business Summit between 23 November and 8 December 2017. A total of 865 pieces of news media mentioned the business event, 156 of which were in national media and 709 in international media. The publications carrying the stories were based in 39 different countries.

Of the 156 press releases appearing in national media, 56 were in digital format, 43 on television, 34 were broadcast on radio and 23 were published in newspapers. International media generated 709 articles, all of which were digital. This data was supplied by the Meltwater platform.

The impact of this level of media coverage is considered to be significant. The ROI for domestic media (within Uruguay) was calculated to be USD 137,006. The international media coverage (709 articles published by international portals) was valued at USD 354,000. This means the total ROI was USD 491,006. The three countries with most publications (outside of Uruguay) were:

- The United States of America: 143
- Spain: 140
- China: 130.

Other countries with a significant level of media publications were: Colombia: 47, Venezuela: 44 and Mexico: 42.

During the course of the business event (30 November to 2 December), the official hashtag: #ChinaLAC2017 was used 3300 times. Other hashtags used were: #ChinaLAC, used 42 times, and China-LAC, used 534 times. (Private Twitter accounts are not included in this analysis.)

From 13 November to 12 December, the hashtag #ChinaLAC2017 was used 4288 times, #ChinaLAC was used 600 times and #China-LAC 1059 times, totalling 5947 mentions of the business event through these three specific hashtags.

### *Impacts on the local economy*

Following the summit, the direct economic impact on the local economy was estimated. This considered the direct expenditure by organisers and delegates in a number of different business activity areas. (These figures are in $US):

- Venue hire: $152,761;
- Hotel and accommodation expenditure: $463,320;
- Audiovisual equipment at the main venue: $274,622;
- Food and beverages at the main venue: $55,881;
- Other: $185,076.

This totals more than $US1,100,000 and represents a significant stimulus for the local economy. It does not take into account the multiplier effect of those funds as they trickle throughout the Uruguayan economy, nor individual spending on things like entertainment and shopping.

## Discussion and conclusions

This case study demonstrates the range of benefits realised as a result of the 2017 China-LAC Business Summit in Uruguay. It highlights the ways in which the business event was used to generate multiple legacy impacts, which will have long-lasting implications for the host destination and the broader Uruguayan economy. It also found that important trade and investment relationships are being forged as a result of this annual business event. These legacies relate to trade, commerce, and the generation of business opportunities for organisations from the LAC region and China.

As stated throughout the case study, however, collation of important and relevant primary statistical data which could more convincingly support the case study assumptions was not possible within the research parameters. The study, therefore, should be seen as a first step toward more thorough investigation of the topic and an important contribution around which future research might be framed. It should also be noted that many of the legacies of this business event will take time to be realised, and it will most likely only be with the benefit of hindsight that the event will be viewed as a major catalyst for economic development and expansion of trade for the countries involved. Another complicating factor is the sensitivity and secrecy which surrounds many of the partnerships made. Future research will need to account for these confidentiality factors.

That understood, in addition to having an outstanding turnout and creating many productive interactions, a number of other key outcomes occured as a result of the 11th China-LAC Business Summit. When Uruguay was announced as the next destination for the summit, a series of agreements were made which led to a change in relations between the two countries. They progressed from having a bilateral commercial relationship to being strategic partners, and the basis for a future free trade agreement was set.

This, along with other information revealed in the study, demonstrates how high-profile business events can be instrumental in creating significant positive change (such as around international trade and relations) if potential legacies are identified and leveraged for in advance. Whilst some may perceive the investment required to organise such a business event as too great, this case study demonstrates the significant advantages to be accrued via such an undertaking when specific important outcomes are planned for strategically.

**Photo 9.2:** Networking at the 11th China-LAC Business Summit 2017

The study was able to measure ROI in terms of publicity generated for the host destination (via social and traditional media exposure) as a result of the business event. This was of great importance to the host city, which could never have invested such financial resources into promotion of their destination. For Uruguay, Punta del Este itself, and the regional tourism industry, this was very significant, given the country is trying to position the city as a leading international business events destination. In so doing, they hope to bring education and business opportunities to the area, and to generate local economic contributions year round.

9

Finally, the summit's organisers have now recognised the importance of documenting activities before, during and after such events, and planning for legacy. They have stated they plan to develop a system for monitoring future business events in order to demonstrate the level of actual and/or potential ROI generated for the host region and for trade relations.

## Implications for future research

Given the importance of this particular business event (and others) for Uruguay's economy, trade relations and development, the Punta del Este Convention & Exhibition Center has now initiated conversations on potential partnerships with the Universidad Católica del Uruguay and the Universidad de la Empresa. The purpose of these partnerships is to work together across different business events, particularly those likely to have significance for the broader industry in terms of their legacies. This will provide important information that can be used by those working in the business events industry, in education and research, and by the Uruguayan community and local organisations. Such research partnerships will help measure whether the objectives of business events (over the short, medium and longer terms) are being achieved, and how. It will also help monitor other kinds of impacts such business events may have in the community and the country more broadly. As such, there is an ongoing discussion with the Universidad de la Empresa about creating an observatory for this purpose.

# Theoretical discussion

Much effort has been devoted to gaining a better understanding of business events' economic impacts (e.g. Dwyer *et al.*, 2000; Jago & Dwyer, 2006; Jago, 2012), which are now readily understood and measured through direct expenditure, value added and employment growth. Their support to government to foster trade, encourage investment, and stimulate urban growth and development has also been acknowledged and investigated. Yet, given their less tangible nature, contributions of business events to foreign affairs and international trade through the development and growth of soft diplomacy have received only limited research attention. However, as demonstrated by this case study, the communication of a country's foreign policy, political values and culture via business events can strengthen the strategic development of soft diplomacy at national, state and local levels.

The history of soft diplomatic engagement is probably as old as humankind. Murray (2017) points to evidence suggesting that Indigenous Australians may have greeted visitors with paintings of star charts and hunting techniques more than 30,000 years ago. The Byzantines practised a diplomacy of "hospitality and wonderment" to conceal their military weakness, allowing them to survive for nearly 1000 years next to neighbours with greater military capacity (Murray, 2017). Such examples are indicative of soft power and are intuitive to humans, who have practised such forms of diplomacy for millennia. Today, the demonstration and application of soft power is complicated due to the range of actors involved, such as institutions and government agencies, and the expansion of communicative domains, both face-to-face and online. Engaging with temporality is a further challenge.

On the one hand, soft power and soft diplomacy require cultivation of long-term relationships built on trust. However, political cycles, changes in government and an emphasis on quantifiable outcomes generate a tendency to focus on economic benefits that can be easily measured, rather than on diplomatic influence where outcomes require longer time horizons for benefits to be perceived and tangible outcomes to materialise. Technological advancement and pace of change also contribute to challenging priorities for diplomatic policies as the internet age magnifies the "paradox of plenty" (Nye, 2008: 100). These challenges require central governments globally to reconsider their soft diplomatic strategy, questioning who is involved and, equally importantly, how can outcomes be measured.

## Soft power

Since the 1980s, contemporary understandings of soft power have been heavily influenced by the work of Joseph Nye. Nye (2008: 94) defines soft power as "the ability to affect others to obtain the outcomes one wants through attraction rather than coercion or payment". According to Nye (2011), a country's culture, its political values and its foreign policies represent its key soft power assets. These resources provide the capacity to persuade other nations to willingly adopt the same goals. Nations hold a soft power advantage (i) when their culture and values match prevailing global norms; (ii) when a nation has greater access to multiple communication channels that can influence how issues are framed in global news media; and (iii) when a country's credibility is enhanced by domestic and international behaviour (Nye, 2008: 96; Snow, 2009: 4).

Consequently, numerous terminologies are used in connection with soft power, and at times overlap, as follows:

- **Public diplomacy** – an instrument that governments use to mobilise resources to communicate with and attract the publics of other countries, rather than merely their governments (Nye 2008: 95)
- **Cultural diplomacy** – the process of mobilising cultural assets as attractors (Singh & MacDonald, 2017)
- **Cultural relations** – reciprocal, non-coercive transnational interactions between two or more cultures, encompassing a range of activities that are conducted both by state and non-state actors within the space of cultural and civil society. The overall outcomes of cultural relations are greater connectivity, better mutual understanding, more and deeper relationships, mutually beneficial transactions and enhanced sustainable dialogue

between states, peoples, non-state actors and cultures. Cultural relations differ from cultural diplomacy in their methodology, objectives and outcomes (Rivera, 2015).

Many scholars have built on Nye's original work, including Pamment (2014) who suggests "the notion of soft power is well-established for analysing the intersections between communication and influence in an international context ... [and is] a broad and passive indicator of the overall attractiveness of a nation, demonstrating the pool of attractive assets that may potentially be leveraged in support of policy goals" (Pamment, 2014: 52). Thus, soft power, in an international context, may be understood as the ability to affect others through attraction, communication and influence to support policy goals.

In essence, soft power of a country rests primarily on its:

- Culture (in places where it is attractive to others);
- Political values (when it lives up to them at home and abroad);
- Foreign policies (when they are seen as legitimate and having moral authority).

While success in promoting these resources rests on:

- Daily communication (regularity of messaging);
- Strategic communication (selecting the right approach to disseminate a country's message);
- Lasting relationships (building and maintaining bonds and friendships).

Given this, it is not surprising that business events, with their focus on, and capacity to facilitate communication and relation-building, can play a central role in the deployment of a country's soft power. It may take the form of a country bidding strategically to host a wide range of high-profile business events in areas of strategic importance for its development, with these events themselves and surrounding publicity offering multiple opportunities for effective messaging in relation to trade, politics and/or culture. At the same time, a business event allows for strategically creating a network of local 'movers and shakers' that is closely connected with influential international ones. Alternatively, countries may initiate, host and participate in regular business events as part of bilateral or multilateral partnerships they have brokered, something China has successfully accomplished over the past two decades as part of its engagement with African and Latin American countries, in the form of its *Forum on China–Africa Cooperation (FOCAC)* and the *China-LAC Business Summit* respectively. Such business events may also provide impetus to further broaden relationships among countries beyond trade, by focusing on strengthening cultural ties.

The following section provides some insights into how China has used soft diplomacy in key areas, often drawing on events, both general and business-oriented, to achieve specific outcomes. Its soft diplomacy efforts are reflected in various rankings. For example, China ranked 2nd out of 20 nations in the Elcano Global Presence Report for 2018 by the Elcano Royal Institute, while it was ranked 8th in the Culture category and 10th in the Engagement category in the Soft Power 30 index in 2019, published by Portland Communications and the USC Center on Public Diplomacy.

## Pillars for developing relationships/partnerships

Economic and cultural spheres are both important pillars for the deployment of soft power to develop effective relationships.

### *Economics*

China has made extensive use of soft power in its economic development. The country's 'economic miracle' over the past few decades has proven attractive to many developing countries. Consequently, the Chinese government has focused on this aspect of its soft power, emphasising its achievements in economic growth and poverty reduction as a blueprint other developing countries could follow (Kurlantzick, 2007: 57). Diplomatic and economic relations were established with a variety of developing countries, primarily in Africa and Latin America, with a view to exploring 'win-win cooperation' agreements with them, using its growing economic power, not only as a trade partner but also as a key investor and provider of aid (Leiteritz, 2012).

China was among the top five source countries for Foreign Direct Investment in 2020 (Leiteritz, 2012; OECD, 2020). Official Chinese government packages for developing countries offer trade, investment and aid, typically without political or economic conditions. Packages are usually related to the building of infrastructure such as roads, rail lines, refineries and ports that directly facilitate exports to and imports from China, as well as the exploration of gas and oil reserves. Chinese companies are then encouraged to trade and invest abroad (Thompson, 2005).

A characteristic instrument of China's economic diplomacy has been business summits, where top-level government officials, diplomats, executives of large companies and investors meet to discuss and explore opportunities for wide-ranging cooperation. Since 2007 there have been four China-Latin America business summits. Similarly, there have been three China-Africa Cooperation Fora since 2006, preceded by two Ministerial Conferences in 2000 and 2006. These high-profile summits are typically the culmination of numerous smaller scale meetings among the various parties aimed at establishing foundations for subsequent summits. Even though it is difficult to assess the overall effectiveness of these meetings in terms of any deals arising from them, the increasing number of participants speaks of their importance as an institutional platform for improved business relations.



China also exerts soft power through foreign aid and development initiatives, including academic, professional and cultural exchanges, in addition to humanitarian assistance. For example, China has committed to contributing to the development of human resources in developing countries by establishing a fund that is jointly administered and used by various Chinese ministries (Foreign Affairs, Commerce, Education, Science and Technology, Agriculture and Health) in order to train personnel. Thus, tailor-made training programs have been devised, as well as training opportunities for seminars and workshops, in addition to government scholarships for university studies, training professionals in different disciplines. Such programs create enduring bonds between Chinese and foreign institutions and individuals. China's strong influence in developing countries is the result of many years of investment in building relations through aid, trade, and cultural and technical exchange – not just the byproduct of China's recently booming economy and soaring demand for African raw materials.

## Culture

Culture represents one of the most important assets in a country's quest to deploy 'soft power' (Nye, 2012). A national culture that is attractive to other countries can be utilised to establish close bonds with people around the world. Utilising arts and cultural assets has long been considered a viable approach to form stronger connections between states as they can provide a common ground despite language differences (Leiteritz, 2012). Cultural institutes are popular diplomatic tools to stand in for and promote a country's values, disseminate a country's culture and language, and to create a more positive image or perception of the country, including its domestic and foreign policy. Organisations such as the Alliance Française, British Council and the Goethe Institute work to build understanding around culture and language by operating domestically, but most importantly through their international outposts. China established its first Confucius Institute in 2004, offering Chinese language and cultural programs. By 2017, there were 525 Confucius Institutes and 1113 Confucius Classrooms in 146 countries across five continents (Chen & Ha, 2020). At the same time, the enrolment of foreign students in China increased from 36,000 in 2000 to about 500,000 in 2018 (Nye, 2012; ICEF Monitor, 2019), leading to China now being the most popular country in Asia for international students, the leading destination globally for Anglophone African students, and the second most popular education powerhouse in the world (Hooi, 2019). Naturally, all these education initiatives are creating long-term bonds and the basis for increased meeting activity.

While international cultural events have expanded from the international expositions of the 19th century towards the regular biennales, capitals of culture and travelling exhibitions of today, nations have also paid much attention to the fact that popular sports can play an important role in communicating a nation's values. According to Nye (2004), the international exposure and the focus on culture and peaceful values make sport a natural soft power device for countries. Murray and Pigman (2014) add that sport obscures boundaries, opens up more channels of dialogue, and brings together different types of people. For example, the two key players in the current case study, China and Uruguay, not only strengthened their trade relations as a result of the China-LAC Business Summit but also deepened cultural ties specifically relating to soccer. China, having inaugurated the China Cup International Football Championship in 2017, invited Uruguay to participate in the tournament in 2018 and 2019. At the same time, two Uruguayan soccer training bases were established in China recently, with the aim of training China's future soccer talent and enhancing China's level in this field (Haoran, 2020). For countries without a long legacy of globally recognised cultural capital, such initiatives provide opportunities to legitimise their cultural assets on the world stage.

## Lessons learned

This chapter presented a case study and theoretical perspectives that offered insights into ways in which business events could be utilised to leverage a country's soft power to achieve specific trade outcomes. In particular, it illustrated how the China-LAC Business Summit created and established mutually beneficial trade relations between government representatives, investors and business operators from the respective countries. At the same time, it highlighted ways in which the business event acted as a catalyst for regional economic growth and development in Uruguay. The deployment of soft power in the lead-up, during and after high-profile business events may also offer a basis to initiate new (or strengthen existing) cultural ties between countries. Clearly, while business events have typically been thought of as common tools to bring together delegates around common disciplinary areas, their communication and relation-building powers allow them to play a central role in the deployment of a country's soft power. In the case of the China-LAC Business Summit, the business event supported the development of international relations in both economic and cultural spheres.

While the need for destinations to accurately gauge the economic benefits of business events has been well established, there has been little focus on researching links between business events and the effective use of soft power. Indeed, when evaluating a business event's performance and contribution in that regard, it may be fruitful to also consider how it was organised and prepared (Pouliot & Therien 2018: 166). This is critical. As Brun (2016) notes, soft diplomacy can also be advanced through 'back-channels' where informal and unseen negotiation can be instrumental in securing ambitious outcomes. Thus, the processes and outcomes of business event diplomacy can be disaggregated by (1) the time preceding a business event; (2) the business event itself; and (3) impacts following a business event. In each of these stages, there may be opportunities to exercise soft power that business event stakeholders have not yet considered. Thus, in the legacy planning for business events, potential opportunities for and specific outcomes of the strategic deployment of soft power in the various business event stages should be identified by relevant representatives of business events destinations.

## References

Alvez, M. J., Sinigaglia, I., Lourdes, A. and Perrier. G. (2019) 11th China-LAC Business Summit 2018, Punta del Este Convention & Exhibition Center, Uruguay XXI and Arquitectura Promocional. https://www.themeetingsindustry.org/web/wp-content/uploads/JMIC_Case-Study_Punta-del-Este_CHINA-LAC-Business-Summit.pdf

Brun, A. (2016) Conference diplomacy: The making of the Paris Agreement, *Politics and Governance* **4** (3), 115-123.

Chen, D. and Ha, W. (2020) Are Confucius Institutes building blocks or stumbling blocks for foreign students in China: An empirical study of 190 countries (1999-2015), *ECNU Review of Education* **3** (2), 235-253.

Dwyer, L., Mellor, R., Mistilis, N. and Mules, T. (2000) A framework for assessing 'tangible' and 'intangible' impacts of events and conventions, *Event Management*, **6** (3), 175-189.

Edwards, D., Foley, C. and Malone, C. (2017) *The Power of Conferences: Stories of Serendipity, Innovation and Driving Social Change*, Sydney: UTS ePress.

Edwards, D., Foley, C. and Hergessel, A. (2016) *Conferences: Catalysts for thriving economies*, report prepared for Business Events Sydney, Sydney: UTS. https://www.besydney.com.au/media/1611/catalysts-for-thriving-economies-research.pdf

Edwards, D., Foley, C. and Schlenker, K. (2011) *Beyond tourism benefits: measuring the social legacies of business events*, report prepared for Business Events Sydney, Sydney: UTS. https://www.besydney.com.au/media/1619/beyond-tourism-benefits.pdf

Fernandez, P. (2017) El País: Buscan alternativa al TLC con China para evitar veto del FA. https://www.elpais.com.uy/informacion/buscan-alternativa-tlc-china- evitar-veto-fa.html

Foley, C., Edwards, D and Hergessel A. (2019) *Direct expenditure attributed to international conferences held in New South Wales 2017-2018*, report prepared for Business Events Sydney, University of Technology Sydney. https://www.besydney.com.au/media/3360/bes002_de6_conferencesinfographic_a4-2.pdf

Foley, C., Schlenker, K., Edwards, D. and Lewis-Smith, L. (2013) Determining business event legacies beyond the tourism spend: An Australian case study approach, *Event Management: An International Journal*, **17** (3), 311–322.

Foley, C., Schlenker, K., Edwards, D. and Hayllar, B. (2010) *A scoping study of business events: Beyond tourism benefits*, report prepared for Business Events Sydney, UTS, Sydney. http://hdl.handle.net/10453/16848

Fonplata (2017) BID y FONPLATA concretan paso clave en su alianza estratégica. https://www.fonplata.org/comunicaciones/noticias/item/2674-bid-y-fonplata- concretan-paso-clave-en-su-alianza-estrategica.html

Haoran, S. (2020) Uruguayan sports chief talks about bilateral cooperation in fitness, training, and more, *Global Times*. https://www.globaltimes.cn/content/1175735.shtml

Hooi, A. (2019) Country in top ranks hosting foreign students, *China Daily Global*, Updated: 2019-12-13 09:11.

ICEF Monitor (2019) China's foreign enrolment growth flattened out in 2018, *ICEF Monitor*, accessed 23 February 2021. https://monitor.icef.com/2019/04/chinas-foreign-enrolment-growth-flattened-out-in-2018/

Jago, L. (2012) *The value of business events*, report prepared for Tourism & Business Events International, Melbourne.

Jago, L. and Dwyer, L. (2006) *Economic Evaluation of Special Events: A Practitioner's Guide*, Altona, Australia: Common Ground Publishing.

Kurlantzick, J. (2007) *Charm Offensive: How China's Soft Power is Transforming the World*, New Haven: Yale University Press.

Leiteritz, R. (2012) China and Latin America: A marriage made in heaven?, *Colombia International*, **75**, 49-81.

Melguizo, A. (2017) It's time for a new era of China-Latin America business relations, paper presented at the World Economic Forum, 27 March 2017. https://www.weforum.org/agenda/2017/03/it-s-time-for-a-new-era-of-china-latin-america- business-relations

Murray, S. (2017) Sports diplomacy in the Australian context: Theory into strategy, *Politics & Policy*, **45** (5), 841-861.

Murray, S. and Pigman, G. (2014) Mapping the relationship between international sport and diplomacy, *Sport in Society*, **17** (9), 1098-1118.

Nye, J. (2004) *Soft Power: The Means to Success in World Politics*, New York: Public Affairs.

Nye, J. (2008) Public diplomacy and soft power, *The Annals of the American Academy of Political and Social Science*, **616** (1), 94-109.

Nye, J. (2011), *The Future of Power*, New York: Public Affairs.

Nye, J. (2012) Why China is weak on soft power, *New York Times*, 17 January 2012. https://www.nytimes.com/2012/01/18/opinion/why-china-is-weak-on-soft-power.html

OECD (2020) FDI in figures, accessed 23 February 2021. http://www.oecd.org/investment/FDI-in-Figures-April-2020.pdf

Pamment, J. (2014) Articulating influence: Toward a research agenda for interpreting the evaluation of soft power, public diplomacy and nation brands, *Public Relations Review*, **40** (1), 50-59.

Pouliot, V. and Thérien, J. (2018) Global governance in practice, *Global Policy*, **9** (2), 163-172.

Punta del Este Convention & Exhibition Center (2017) *Punta del Este Convention & Exhibition Center Business Report 2016-2017*.

Rivera, T. (2015) *Distinguishing Cultural Relations from Cultural Diplomacy: The British Council's Relationship with Her Majesty's Government*, Los Angeles, USC Center on Public Diplomacy at the Annenberg School.

Singh, J. and MacDonald, S. (2017) *Soft power today – measuring the influences and effects*, report prepared for the British Council.

Snow, N. (2009) Rethinking public diplomacy, in N. Snow and P. Taylor (eds), *Routledge Handbook for Public Diplomacy*, New York: Routledge.

Thompson, D. (2005). China's Soft Power in Africa: From the "Beijing Consensus" to Health Diplomacy. *China Brief: a Journal of Analysis and Information*, **5** (21): 1-4.

Uruguay XXI (2017) The China-LAC 2017 summit was the largest summit among all Latin American editions, accessed 14 December 2017. https://www.uruguayxxi.gub.uy/en/news/article/the-china-lac-2017-summit-was-the-largest-summit-among-all-latin-american-editions/

# 10 Leveraging Legacy

> *After a year of great challenges and difficulties for our industry, the research about the value of MICE events, their direct impact and legacy for their communities and destinations is becoming more relevant today than ever.*
>
> Alejandro Abulafia, Executive Manager, Punta del Este Convention & Exhibition Center

The highly infectious nature of COVID-19 meant almost every part of the world experienced some sort of seclusion as a result of preventive lockdown measures that began in 2020. In this context of limited mobility within and across countries, business events were cancelled or rescheduled (Achakulvisut *et al.*, 2020; Viglione, 2020), leaving hundreds of thousands of people who had previously derived their income via roles connected to the staging of business events (directly or indirectly) out of work. At the same time, millions of industry associations, researchers, businesses and delegates around the world have been left without access to the many benefits of conference, exhibition and trade show attendance.

As the business event sector confronts the challenges which lie ahead, the case studies in this book remind us of the power such meetings have to stimulate economies, generate knowledge, foster collaborative relationships, drive innovation and support job growth. They are powerful tools for educating, connecting, provoking thought and driving change (Foley *et al.*, 2013; Foley *et al.*, 2014; Edwards *et al.*, 2017). Beyond this, the case studies and theoretical considerations presented demonstrate the potential business events hold to contribute to such things as achievement of global sustainability initiatives, implementation of important soft diplomacy strategies, creation of knowledge spirals, with impacts reaching well beyond the event itself, and the establishment of specialist industry hubs, or nodes, in host regions. As our understanding of the full potential of business events to create such legacies increases, such outcomes are being planned for, and are beginning to be documented; an important objective of this text.

Whilst most people acknowledge that business events generate connections and enable knowledge creation and sharing, the case studies presented in this text demonstrate that business event organisers are becoming *intentional* about creating positive, long-term legacy outcomes, and it is this *intentionality* which has the potential to increase the value of the industry. Networking already occurs during business events because like-minded individuals are gathered in the one place discussing cutting-edge concepts, but how many *more* networks could be formed, or how much *stronger* might these networks be, if they had been strategically planned from the outset? Business events already create partnerships and collaborations because of the shared social spaces and opportunities participants experience, along with the innovative attitudes those experiences inspire. But how much *more* successful or common might those collaborations be if there was intention and planning around their inception? Diplomatic affairs, foreign trade relations and local research environments or economies already benefit from conference activities because of the knowledge and information being shared. Still, the question remains: just how much *more* powerful might those benefits have been if event designers had been *intentional* about ensuring such legacies were created, managed and monitored? Finally, how much more support might public and private entities provide for business events if the industry as a whole was able to initiate systemic changes and invest in the development of tools, processes and practices which encourage, support, measure and leverage the legacies generated?

This concluding chapter reflects on the success factors discussed throughout the book, the ways in which business event legacies were leveraged and the challenges and barriers encountered in the process of planning for and achieving those legacies. It also offers some suggestions on ways the sector might seek to better plan for, evaluate and formally report on legacy, and lessons learned from each chapter.

## How legacies were leveraged in the case studies

The business events presented throughout this text had a number of things in common, not least of which was their clear intention at the outset to achieve identified legacy objectives. The Business Events Sydney (Chapter 2) and ICC Sydney (Chapter 6) studies differed slightly from the other case studies. Chapter 2 identified common legacy outcomes across a range of different conference, industry and delegate types, while Chapter 6 detailed the benefits of an overall operational legacy approach by a large convention centre.

Each of the event-specific case studies had certain elements in common. They all had a foundation of at least one particular legacy outcome, they all incorporated innovative ideas into planning for those legacies to occur, the linking of local and global networks were priorities in each case, and all gave consideration to conducting research to document their legacy outcomes. Together, the cases provide a solid foundation upon which to build greater understanding of how to create legacies via business events.

## Intended legacy outcomes were clear at the start

Each business event presented in the case studies was designed to generate legacies of much greater significance than direct tourism expenditure. The Sarawak event (Chapter 3) had multiple intended legacy objectives (assisting the goal for Malaysia to become a developed nation by 2020; promoting Sarawak as an international meetings destination; upskilling the Malaysian business events industry; increasing awareness about the plight of the orangutans; and supporting Malaysian children's education), as did London Tech Week (Chapter 4) with its ambitious plans to (amongst other things) place London at the tech heart of Europe. Knowledge generation for those working in heart health across Europe was a very significant planned legacy outcome of the EuroHeart Care Conference (Chapter 5) and ICC Sydney's Feeding Your Performance program had a wide range of legacy objectives including supporting local industries, contributing to regional economic development and generating social capital. The Fintech Corner at the SIBOS conference (Chapter 8) aimed to drive collaborations between small local startups and larger, well-resourced global players to support the Geneva fintech industry, while Chapter 9 was very clear that trade, the exertion of diplomatic relations and stimulating the ongoing off-season economy in Punta del Este (by building a local business events industry) were driving motivators for the event. Essentially, each of the events studied exemplified the notion that there are many more significant legacy outcomes deliverable via business events than tourism expenditure. The legacies were strategically planned for and initiatives were put in place to ensure they would be achieved.

## Intended legacy outcomes involved or benefited a range of stakeholders

Most of the events discussed in the text made reference to a range of stakeholders who would benefit from the intended legacy outcomes. Whilst the scope of the research was in some cases limited, almost all chapters referred to multiple stakeholder groups being involved in the planning stage and working together to create successful outcomes. Examples of collaborations between event organisers and local suppliers can be found in ICC Sydney (Chapter 6), which thoroughly documents stakeholder groups involved in the Feeding Your Performance program and the ways in which they benefited from legacy creation; Malaysia (Chapter 3), which discusses collaborations between convention bureaus, tourism boards and local organisations; and Punta-del-Est (Chapter 4), a collaboration of business event organisers and government agencies and associations.

## Innovative ideas

One of the success factors evident throughout the case studies was the incorporation of innovative ideas into event design. Chapter 3, for example, incorporated some innovative social events into the conference program designed to achieve legacy objectives (delegates bringing books; visiting the orangutans; 'painting the town green'). These events aimed to form a unique experience for event delegates and hence increase the competitiveness of the destination as a business event host.

Another example is presented in Chapter 6, which details ICC Sydney's strategy of providing balanced nutritious meals for business event delegates with ingredients sourced from local suppliers. One of the reasons for choosing this approach was to increase 'the performance' of the delegates, while offering quality locally sourced food as a point of differentiation from other venues. The approach highlights the potential of convention centres to deliver benefits to the region in which they are located by supporting agriculture and local farming through sustainable procurement via short food supply chains. ICC Sydney's procurement steps focused on reducing the purchase of non-sustainable foods, building collaboration with suppliers, reducing food miles and corresponding positive environmental impacts. Key to this was sourcing foods that were produced by following sustainable regulations. These initiatives not only brought benefits to local economies and communities, but were used to improve ICC Sydney's visibility as a marketing point of difference to set the venue apart, and to demonstrate best practice for food procurement to the sector.

## Linking local and global

The business events in the case studies were all attended by domestic and international stakeholders (e.g. delegates, associations), which created local and global linkages in two ways. First, the local talent, sector strength and research capabilities of the event location were showcased to a global audience, who gained innovative ideas by accessing local knowledge and forming collaborations with local delegates (as evident in Chapters 2, 4, 5, 7 and 9). Second, by engaging the community (such as local suppliers or firms), local products and services were consumed by a global audience (Chapters 3, 6, 8 and 9). These local-global networks and collaborations were vitally important for stakeholders in gaining credibility.

The local-global element is also conducive to soft diplomacy wherein business events can be utilised for multilateral partnerships (Chapter 9). Soft diplomacy through business events can be achieved by fostering national foreign policies, political values, cultural understanding, and international cooperation and business, via the activities conducted pre, during and post the business event.

## Research initiatives

Whilst much of the research undertaken in preparing the case studies for this text could be further developed, each one makes a useful contribution towards a better understanding of how business event legacy outcomes can be planned for and evaluated. Business events are often major undertakings that require intense planning to ensure the scheduled activities are delivered professionally, safely and successfully. There are often tight time-frames and limited resources, and the pressure of coordinating such a dynamic and complex interaction of activities can be highly challenging. As a result, things such as research, legacy measurement and evaluation can be secondary considerations, resulting in lost opportunities to demonstrate to various stakeholders the value created, and to leverage that value for future events.

Incorporating research initiatives into the planning and evaluation of business events must become an integral component of professional practice. Human and financial resource allocations can be incorporated into initial budgetary and planning stages in order that outcomes can be tracked, achievements can be communicated, and lessons can be shared and learned. This will not only improve the ways in which events are delivered, but will enable legacy creation to be maximised and the value proposition of individual events and the industry, as a whole, to be realised.

## Challenges faced in leveraging legacies

A number of challenges and barriers to success were referred to in the case studies. Table 10.1 presents the barriers in four time periods: those occurring before, during and after the business event, and those of no specified timeframe.

**Table 10.1:** Barriers and challenges to hosting business events

| Timeframe | Barriers | Stakeholders affected | Ch |
|---|---|---|---|
| Before the business event | Regulatory challenges<br>Collaboration among organisers<br>Organisational challenges (economic risk, coherence of goals, existence of measures)<br>Personal challenges (motivation of individual researchers) | Hosting organisations (professional business event organisers, university) | 7 |
| During the business event | Infrastructure<br>Lack of preparation<br>Poor coordination of the business event (exhibition space, media coverage, participation of attendees)<br>Lack of a media strategy | Conference organisers (professional business event organisers, convention centre)<br>Further implications for delegates | 8, 4 |
| After the business event – to gain long- term benefits | Mismanagement of long-term visibility of the event<br>Funding or communication necessary for long-term research projects<br>Differences among stakeholders<br>Lack of a business event long-term strategy formed by conference organisers<br>Lack of follow-up measures | Delegates<br>Further implications for business event organisers, government and others | 2, 7, 8 |
| Barriers with no specific timeframe | Absence of logistics strategy<br>Contracts management<br>Poor inter-organisational communication<br>Climate | Convention centre<br>Suppliers' value chain (farmers, suppliers, producers & distributors) | 6 |

Barriers which occur prior to the business event were mentioned in Chapter 7. These barriers involved collaborative issues among co-organisers, such as a scientific society and a university, and conflicts regarding the sharing of responsibilities for business event planning and organisation. Moreover, organisational challenges involved insufficient guarantees in the case of a business event ending with a deficit, the business event not aligning with the university's goals, and/or a lack of measures for delivering a successful event.

Chapter 7 also found that the motivation of academics to participate in organising a business event was often missing, as academics are predominently evaluated for their teaching and research activities. Further, regulatory challenges in resolving conflicts of interest among practitioners and scientists might act as a barrier for business event legacy outcomes.

Barriers during the business event are characterised as challenges perceived by business event organisers, but their occurrence has implications for business event delegates. To illustrate, Chapter 4 mentioned several challenges, such as problems with infrastructure (hotels and other tourism value chain actors), and management problems such as lack of preparation and coordination of event activities. Moreover, Chapter 4 reported participating firms complaining about mismanagement of media coverage by organisers as media attention was focused primarily on one firm.

Other barriers were identified as impeding the establishment of long-term benefits. These barriers included the mismanagement of the business event visibility (Chapter 4) resulting in a lack of interest from wider stakeholder groups; difficulty in managing various stakeholders (different sectors, country of origin, nature of the research); issues in communication among collaborating researchers; lack of funding, which hampered benefits initially formed during the conference (Chapter 2); and the lack of a long-term strategy, including a failure to establish follow-up measures by business event organisers (Chapter 7).

## Designing innovative business events for legacy creation

Event management implies intentional management or design of 'what happens' at an event, in order to achieve certain outcomes, and as Berridge (2007) states:

- All event environments are created;
- All experiences within event environments are purposefully designed;
- All stakeholders are the direct recipients of the designed experiences.

Accordingly, business event management might be conceptualised as a creative product; a staged, temporary and intentionally designed experience in which multiple actors interact, share, create, contribute, serve, explore, learn and connect to achieve a range of planned and unplanned outcomes.

Under this construct, business event organisers can be perceived as designers, and as such need to think from a larger design perspective. What is the final product to be achieved? Who might use the product? How can it best meet the needs of the user? What barriers, challenges or shortcomings might various design options have that other options don't have? How can the event design be improved to create even better outcomes and more significant legacies in the future? And how can the outcomes of all the design efforts be captured and made tangible/transferable/reportable in order that the value can be leveraged?

The extent to which event experiences engage and inspire participants, stimulate their thinking, impact their practices, facilitate future innovation and entrepreneurial or partnering opportunities, or generate solutions to important global issues depends, to a large extent, upon the nature of their design. They require different design layers with numerous components that can be experienced on several levels, and recipients should be able to extrapolate the meanings they are presented with and use them for a range of purposes (Berridge, 2007). The design of events is evolving as more free-thinking and experiential formats, aligned with the delivery of specific objectives, are adopted (Berridge, 2012; Crowther, 2014). As business events can involve multiple and varied actors with common interests, inevitably new information, cutting-edge ideas and pressing issues are discussed. It is not surprising, then, that outcomes such as new business ideas, product innovations, partnerships and collaborations, or steps toward solutions to issues are common eventualities. Event designers can deliberately affect this environment by incorporating aspects of design thinking into their planning stages to provide spaces and activities for both capitalising upon existing opportunities and creating new ones (Crowther et al., 2018).

## Design thinking

Design thinking is a human-centred approach aimed at creating innovative solutions in response to wicked problems. The focus is on how humans might interact with/use/feel about a designed product (extremely user-centric) and seek to find effective solutions to meet people's needs (Stevens, 2020). Essentially, it is a solution-based approach that focuses on *design* to solve problems. It is an iterative process where what is 'known' is constantly being acquired, questioned and challenged to help redefine the problem and continue to evolve alternate and improved strategies and solutions (Dam & Siang, 2020), which may not have been initially apparent. It is about developing new ways of thinking, based on what was previously known and what was previously done – but improving them to create and access new knowledge (how did the 'consumer' react to the previous 'product') and based on that new knowledge, doing things in new ways. It has long been used by artists (in literature, music and visual arts), scientists and (more recently) in business (marketing and product design) for practical and creative problem-solving. Based on this understanding, it is evident that people can have a natural aptitude for design thinking. The attributes design thinkers use are:

- **Empathy**: ability to imagine the world from multiple perspectives (i.e. colleagues, clients, end users and customers – current and prospective). This means taking a 'people first' approach so you can imagine solutions that are inherently desirable and meet explicit or latent needs.

- **Observation**: can observe the world in minute detail, noticing things others do and don't do and using these insights to inspire innovation.
- **Integrative thinking**: ability to see all the salient (and sometimes contradictory) aspects of a confounding problem and create novel solutions that go beyond and dramatically improve existing alternatives.
- **Analytical process thinking**: producing 'either/or' choices.
- **Optimism**: assuming no matter how challenging the constraints of a given problem, at least one potential solution is better than the existing alternative.
- **Experimentalism**: rather than just relying on incremental tweaks, design thinkers pose questions and explore constraints in creative ways that proceed in entirely new directions.
- **Collaboration**: given the increasing complexity of products, services and experiences, we no longer hold the myth of the lone creative genius working in isolation. We now have enthusiastic interdisciplinary collaborators. The best design thinkers don't simply work alongside other disciplines; many have skills in a number of areas, collaborate widely and integrate perspectives

(Harvard Business Review *et al.*, 2020: 13).

Great design also includes aesthetics; how something appeals emotionally. As basic needs are met, sophisticated experiences that are emotionally satisfying and meaningful are required. Such experiences are complex combinations of products, services, spaces and information. Design thinking is a tool for imagining these experiences as well as giving them a desirable form (Harvard Business Review *et al.*, 2020). Whilst there is significant literature on design thinking, the aspects we consider most relevant to business event design are:

- Focus on the user;
- Ongoing testing, feedback and improvement;
- Tackling wicked problems.

## Focus on the user

According to Dam and Siang (2020), design thinking taps into capacities we all have, but are often overlooked when we focus on conventional problem-solving processes. Rather than focusing purely on analytical or rational approaches to issues (i.e. thinking about problems using previously held patterns of thought, knowledge or attitudes), design thinking adds innovation by considering things like pattern recognition, intuition and the ability to construct ideas that have emotional meaning, as well as functionality and recognition that concepts, feelings or ideas can be expressed in ways other than via words or symbols (Dam & Siang, 2020). Because it looks at things like emotions, needs, motivations and drivers of behaviour, it is more sensitive to and interested in the context in which users are going to interact with the product and the feelings, challenges or obstacles they may experience during that interaction, than with the product itself (Brown, 2008; Dam & Siang, 2020). It takes the original problem as a suggestion, not as a final statement, and considers what the real and underlying issues are. By combining rational approaches with holistic, user-centred perspectives, design thinking is said to be a way of matching human needs with available technical resources and practical constraints (Brown, 2008; Dam & Siang 2020). Rather than thinking 'outside the square', it is thinking that is 'at the edges of the square' where the two ideologies are combined to form an integrated approach to complex problems (Dam & Siang, 2020).

As designers of business events, practitioners should consider the needs of different event stakeholders in the early stages of planning. Considering such needs will involve consultation with those stakeholders and a genuine effort to understand what outcomes they hope to achieve. Chapter 7, for example, noted reasons for attendance and non-attendance by early career researchers. This knowledge can provide a useful starting point for further research which may be of use to business event organisers. Similarly, if an event organiser intends to generate a certain legacy outcome, design thinking might encourage them to consult with intended legacy recipients at the outset.

## Ongoing testing, feedback and improvement

Business events are dynamic, operating in a context in which circumstances and conditions are constantly changing and community attitudes and values are constantly shifting. Hence, formulaic approaches to event design will not work in all situations.

Design thinking is a non-linear process wherein designers continuously use results to review, question and improve their earlier assumptions, understandings and outcomes (Brown, 2008; Dam & Siang, 2020; Harvard Business Review *et al.*, 2020). Results from the final stage of the initial work process inform understanding of the problem, help to determine the parameters of the problem, enable redefinition of the problem, and provide new insights that can lead to alternative (and even better) solutions (Dam & Siang, 2020). Advocates of design thinking suggest tools such as brainstorming, ongoing experimentation, prototyping (trialling those possible solutions) and sketching (to give visual representation of ideas) assist the design process (Liedtka, 2018; Dam & Siang, 2020).

Business event professionals can use these tools to generate ideas and test them out on volunteers or in small group settings. As events progress, mechanisms can be put in place to enable ongoing evaluation and to provide opportunities wherein feedback is actively used to help participants co-create their own experiences. As is the case with wicked problems, the possibilities are limited only by one's imagination and constant evolution.

## Tackling wicked problems

The concept of wicked problems put forward in the 1960s by mathematician and teacher, Horst Rittel, encompassed those design problems where information was confusing and where there were numerous (and often conflicting) pieces of information, needs, clients and decision makers to be considered (Buchannan, 1992). Design problems were said to be 'wicked' because design does not have its own special and concrete subject matter, other than what a designer conceives it to be (Buchannan, 1992). In this way, design problems are unable to be fully determined, which is in stark contrast with disciplines like science which deal with principles, laws and rules to be determined or that have been determined. A business event designer must consider a range of stakeholders with diverse needs and objectives, and from different social, cultural, physical and economic environments for each event. Their skills must be fluid and transferable enough to enable them to accommodate competing interests and to modify, monitor and adapt their products as required.

Chapter 3 provides examples of wicked problems that might relate to business event delivery and partnerships for sustainability initiatives: that of how to design events which lead to the establishment of great collaborations and successful partnerships; and that of how to maintain links or affiliation with those partnerships over time in order to track legacy outcomes.

There are countless wicked problems event designers may face every day. Not all will require complex solutions, and it is sometimes enough to have a willingness to challenge assumptions and embedded 'knowledge' in order to have the clarity and vision to be able to see an issue from a new perspective (Dam & Siang, 2020).

Accepting the inherent nature of wicked problems, adopting the user-centred approach and being committed to ongoing evaluation, testing, modification and improvement will be the foundations for improved legacy-creation strategies for event design.

In summary, the concept of design thinking has been included in this text because business event designers are confronted with a range of complex or wicked problems every day, and such strategies will assist them to devise solutions. Importantly, we believe the basic principles of design thinking can assist practitioners to:

- Design events in ways which maximise legacy potential;
- Capture, document and leverage the full extent of the value business events generate.

Events can be viewed by designers as 'wicked problems'; with the 'problems' being the myriad of possibilities the event may hold for creating legacy outcomes, both in the short term and well into the future. Rather than event planning and delivery being viewed as a linear process whereby a long list of sequential tasks are completed in order, it might be seen more as a continuously evolving process that provides opportunities for co-creation of the experience, based on emerging needs and actions of stakeholders. For example, design thinking could be used to design business event programs and spaces that *intentionally* bring together individuals or groups from diverse industry sectors in order to form new and beneficial partnerships. It could be used to create learning opportunities strategically designed to inspire such partnerships, with the potential to develop the solutions needed locally and globally. It could also be used to find ways of maintaining links with those partnerships and monitoring their achievements in order to more accurately measure the full value business events bring.

Similarly, in relation to Chapter 7 (on academic conferences), the empathetic approach encouraged in design thinking could be useful in creating event experiences that better meet the needs of delegates; for example, early career researchers and academics who are unsure of the benefits and opportunities to be gained from attending a business event. Ongoing relationships with delegates could be established to understand how they experienced the event and to track any changes or developments that occurred for them as a result of attendance. Those responses could be used to improve future events and to further demonstrate value outcomes.

Indeed, the legacy issues and examples of the long-tail benefits (Edwards, *et al.*, 2017) accorded by business events and discussed in this text could be thought of from the human-centred, wicked problem perspective, suggested by design thinking, as generating creative and innovative new approaches. Design thinking can facilitate ways of approaching event design to take into account the specific cultural and human-centred contexts in which the business event will take place. This will include the use of testing, trialling, sampling and using creativity, innovation and ongoing research and evaluation to continuously improve and document the experience.

## A way forward: The seven-step framework for business event legacy planning

Despite the challenges, the suite of business event legacies identified in this book are extensive. They challenge us to reconsider the way we value and measure business event legacies, and to expand our understanding and vision of what is possible.

The findings in this book confirm a significant mismatch between the way governments and other stakeholders currently value the business events sector and the legacy outcomes they deliver. Few would dispute the value of tourist dollars to a destination, and business events are known to attract high yield visitors. Yet these chapters demonstrate that contributions go well beyond tourism to knowledge economies, industry sectors, and communities. Leaders in the business events sector understand they are "not travel agents but change agents" (The Iceberg, 2019). However, many political leaders and a large section of the business events industry itself are still in the dark when it comes to understanding what can be leveraged from their business events or their sectors.

The breadth of benefits discussed in this book indicate it is not possible to use a single measurement tool for every business event. Methodology will always need to be tailored to meet the contextual requirements of the individual business event. Recent external shocks to the sector and technological advances have accelerated the move to hybrid conferences, making it more important that legacies beyond tourism continue to be a focus for event organisers. To this objective we present a seven-step model (see Figure 10.1) which may assist business event organisers to maximise legacy. The model builds on a six-step model previously proposed (Foley *et al.*, 2019).

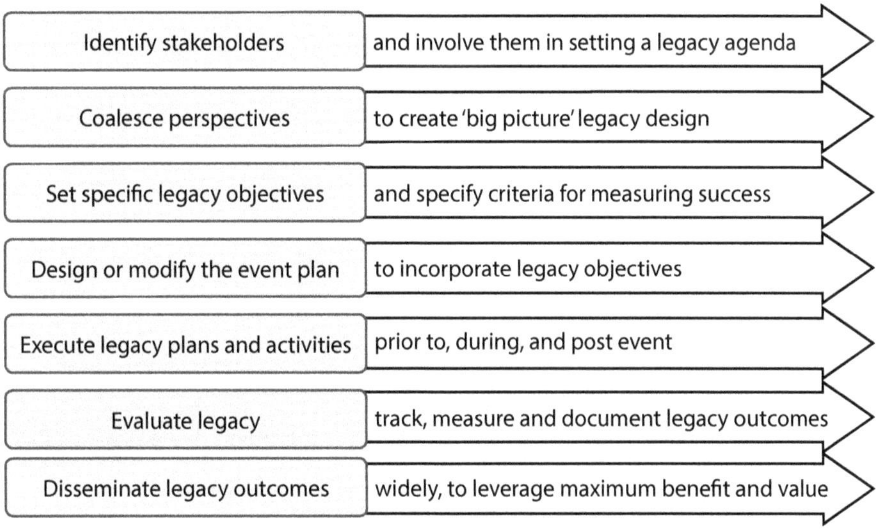


**Figure 10.1:** The Seven-step Framework for Business Event Legacy Planning

1. *Involving stakeholders* in setting legacy agendas may include government, local and global associations, industry leaders within the destination, leading researchers within the destination, conference organisers, local residents or businesses, bureaux and venues (see Chapter 3 for more discussion on stakeholder identification).
2. *Coalescing* the legacy perspectives of key industry and/or community issues and opportunities is important. By identifying and fusing such issues, event organisers will help stakeholders to come together for a common purpose and motivate them to unite behind legacy initiatives.
3. *Specific legacy objectives* need to be set in order to plan for their execution. Whilst many conferences yield legacies regardless of whether they have been planned for, the failure to set objectives can mean lost opportunities for making a difference.

4. *Design* your event to accommodate your legacy objectives. Once objectives have been set, there must be a plan for ensuring the legacies are realised. Planning for the evaluation of legacy outcomes is essential, hence the overall event design should be evaluated and configured (or reconfigured) to accommodate the established legacy objectives.
5. Once all planning is in place, *execute your legacy plans*. This involves actions prior to the event (i.e. establishing relations, putting processes in place, etc.), during the event, and after the event (i.e. follow-ups, tracking processes, etc.)
6. Whilst *evaluation* is universally important, different outcomes call for different methods of evaluation. As a starting point, methods of data collection and analysis are included in this book. Partnering with a research specialist (university or consultant) can fill skill and knowledge gaps.
7. *Dissemination* of what was achieved is important for a number of reasons. Legacy outcomes should be communicated widely. It is important for all stakeholders to understand the full value of the business event and the contribution they may have made. Legacy outcomes should be included in every business event study or evaluative report and then communicated to governments, communities, industries and universities, as well as event organisers, individual delegates and any stakeholders that have invested in the business event.

## Leveraging legacy

In reviewing the chapters of this text, we identified five pathways through which legacies were realised:

- the mutual reinforcement of benefits
- long-term indirect benefits
- development of benefits over time and space
- benefits formed by stakeholder collaboration
- stakeholder collaboration managed by organisers.

These pathways are presented diagrammatically (see Figures 10.2, 10.3, 10.4, 10.5 and 10.6) below.

### *Mutual reinforcement of benefits*

Across the chapters we found some legacy outcomes were leveraged through the effective combination and synergy of benefits in real time (Figure 10.2a) or shortly after (Figure 10.2b). That is, benefits can have a multiplying effect in that they can have a positive impact themselves as well as reinforce other benefits. Such outcomes arise as the result of knowledge interactions and transfer that occur within the contextual environments of the business event, as explained in Chapter 5.

**Figure 10.2:** Mutual reinforcement of benefits

**a:** Enhanced knowledge sharing (during the business event)

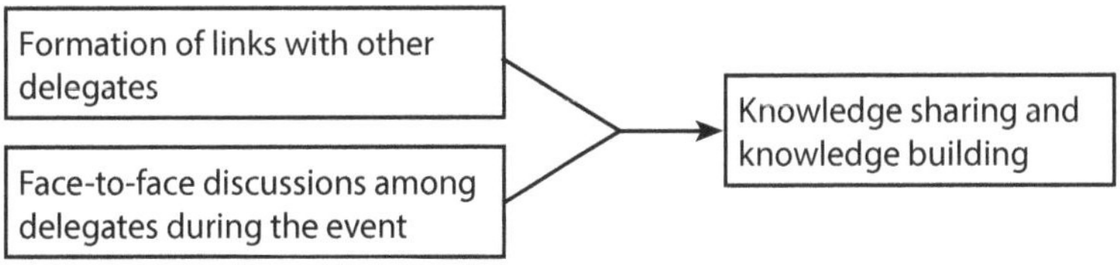


**b:** Enhanced collaborative atmosphere at the workplace (after the business event)

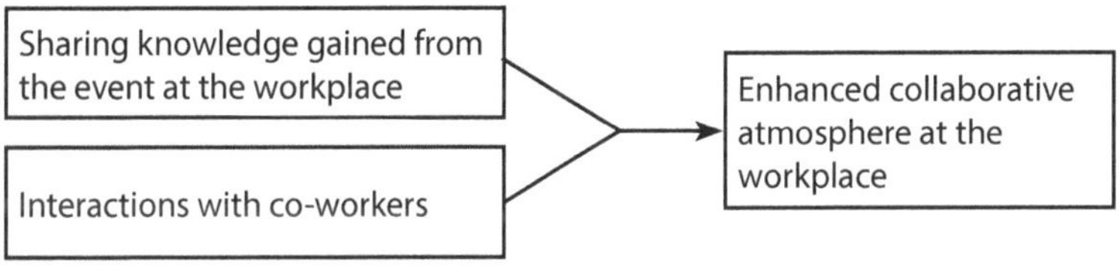


To illustrate, Figure 10.2a portrays the situation where individuals attending the business event form links with other attendees and also engage in intensive face-to-face discussions (evidence of these behaviours can be found in Chapters 7, 5 and 2). As per figure 10.2a, an effective combination of these two already formed benefits from the business event can create a further benefit of extended knowledge sharing and knowledge building among delegates.

Knowledge sharing formed by participation in the business event is presented as a benefit in several studies focusing on legacies for academia and industry. However, this benefit can also be leveraged by effective networking which involves face-to-face communication and discussion among participants. These are not only actively driven by attendees but are also facilitated by the organisers through provision of social space for personal engagement, such as coffee breaks, social events or pre- and post-conference events. Hence, while networking and knowledge sharing are benefits, their combination can form even stronger legacies for knowledge building and enhanced collaboration among attendees.

Another example in Figure 10.2b refers to a legacy described in Chapter 5 (Jonkoping Euro Heart Care), which found that the benefits occurring during and shortly after the business event had a multiplying effect. Following the business event, the nurses returned to the workplace (e.g. hospital department) and shared their knowledge among their co-workers, leading to an additional benefit of increased workplace collaboration.

## *Long-term indirect benefits*

Chapter 2 also presented the model *#Dynamics of Conference Legacy*, outlining a framework for analysing business event legacies over time. This framework incorporated an intrinsic dimension in which immediate and fundamental benefits were obtained by delegates or organisations during the conference; an application dimension in which individuals and organisations implemented practices and knowledge gained at the conference in their work; an attitudinal dimension whereby there was an immediate benefit to the community gathered at the conference; a knowledge economy dimension wherein there were long-term benefits for the community and for the economy at large; and transformative social network dimensions, facilitating short and long-term benefits for those who participated, through to the conference itself and the community at large. These dynamics are visualised in Figures 10.3(a, b, c, d), 10.4 and 10.5(a, b).

For example, in Figure 10.3a, knowledge gained during the business event led to professional development in the long-term, such as in the case of nurses and their knowledge development (Chapter 5). Similarly, academics converted new knowledge into research and publications (see Chapter 7). Moreover, the other example depicted in Figure 10.3b highlights that collaborations formed during the business event led to funding opportunities (Chapter 2). The third example (Figure 10.3c) explains the formation of benefits found in Chapter 7. The authors, in this case, reported that visibility and reputation of host organisation researchers participating in business event organisation led to invitations to speak at other conferences or seminars. Finally, Figure 10.3d demonstrates that there is a link between networking and the realisation of collaboration and investment opportunities (see Chapters 3, 4, 8 and 9).

**Figure 10.3:** Long-term effect of benefits

**a:** Knowledge gain leading to professional development

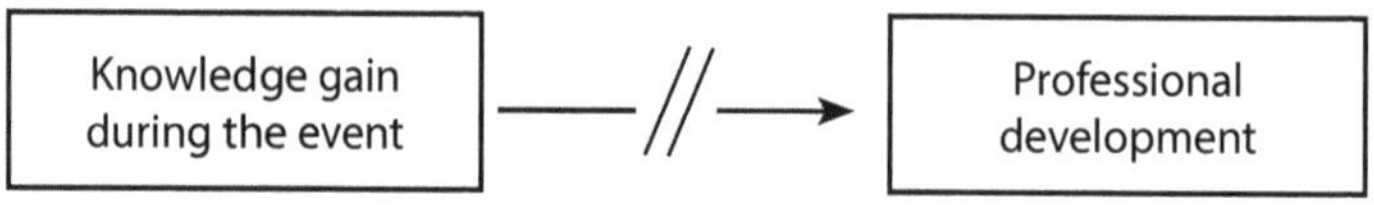


**b:** Collaboration leading to funding

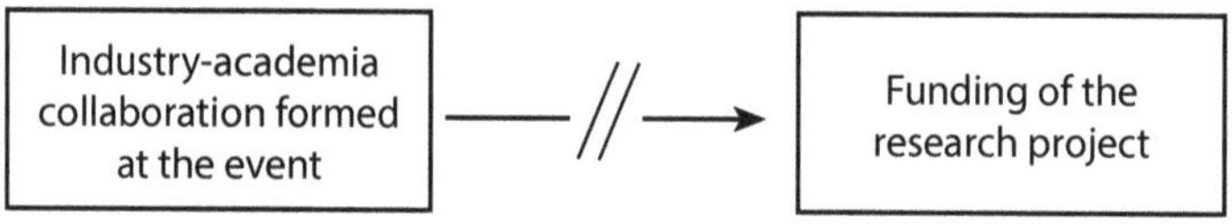


**c:** Visibility and reputation leading to invitations to be a speaker

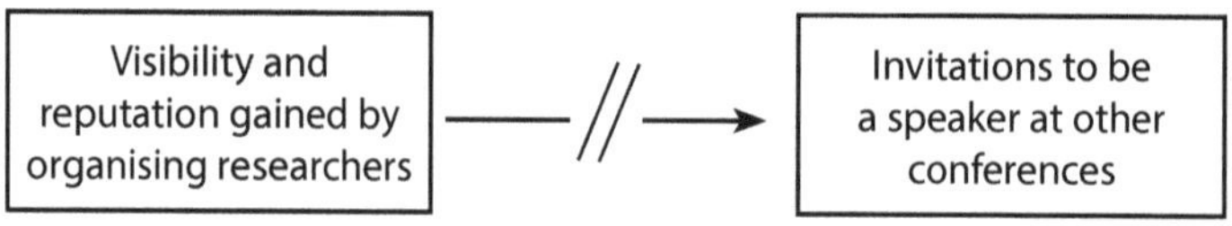


**d:** Networking leading to investment opportunities

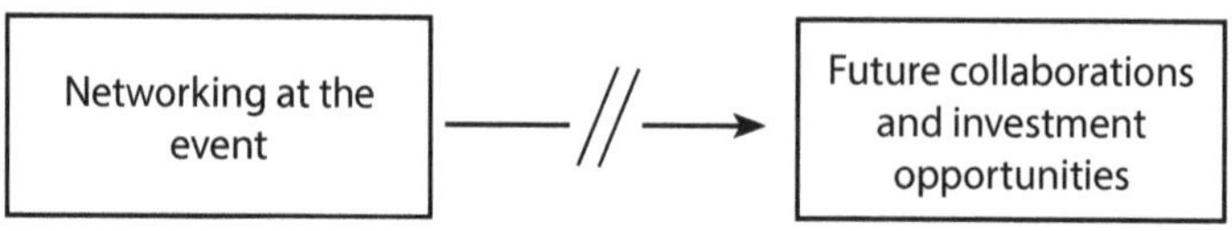


Overall, leveraging long-term benefits is an area in which more research is required, and has been mentioned by several studies as challenging to maintain. To this end, two possible strategies were identified which could facilitate the transformation of newly formed research collaborations into research projects.

First, Chapter 2 suggests that understanding the differences between delegates, such as industry sector, country of origin, career stage and length of professional experience (e.g. early-career or experienced researcher) could provide insights into the diversity of expectations and needs. In doing so, organisers could tailor legacy objectives to better meet identified needs and expectations of various delegates.

Second, the visibility of several business events was impacted by media coverage (see Chapters 3, 4 and 8). Careful management of media coverage is important to facilitate immediate benefits of visibility and reputation of the business event, as well as fostering future benefits such as speaking invitations, improved destination reputation, and building community awareness of the scientific area.

## Development of benefits over time and space

This pathway for leveraging legacy combines the development of benefits (time) and their allocation to various stakeholders (space). This means that the benefits would not only be transformed over short and long periods of time, but multiple parties could be involved directly and/or indirectly at different periods of time. An example (Figure 10.4) drawing on Chapter 6 is used to demonstrate this point.

**Figure 10.4:** Development of benefits over time and space

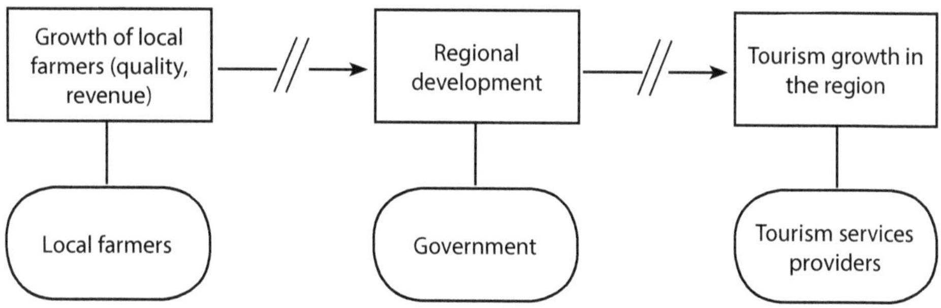


In Chapter 6, the implementation of a nutritional program by a convention centre involved sourcing quality food products from local suppliers and serving nutritious meals to business event delegates. The benefits in this study were realised by local farmers in the form of sales and business growth. In turn, business growth flows to regional development, thereby supporting local and state government agendas for regional development. In the long-term, the business event predicted potential benefits for regional tourism growth arising from increased branding and product recognition. Local wineries linked to the convention centre are already reporting increased patronage from visitors who were introduced to their wines at business events.

## Benefits formed by stakeholder collaboration

10

Another pathway for leveraging benefits is the formation of links and partnerships among various stakeholders. While the previous three pathways focused on the benefits assigned to single stakeholder groups, a fourth pathway reflects the benefits created by their cooperation. Figure 10.5 diagrammatically presents these relationships and their outcomes with a focus on collaboration between two parties (Figure 10.5a) and collaboration among multiple parties (Figure 10.5b).

**Figure 10.5:** Benefits formed by stakeholders' collaboration

**a:** Collaboration between two parties

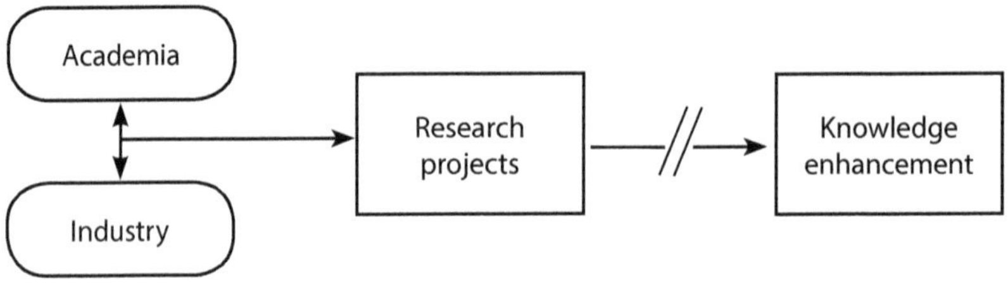


**b:** Collaboration among multiple parties

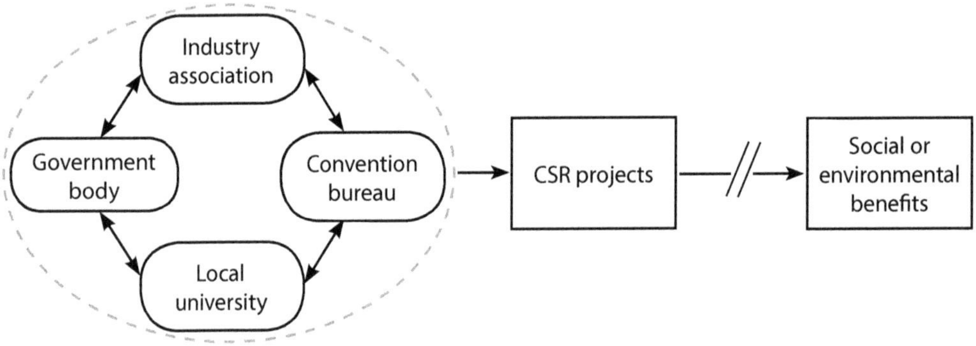


First, the collaboration between two parties is illustrated using academic and industry links (Figure 10.5a). In particular, the cooperation of researchers and industry representatives leads to the formation of research projects that are funded by industry and combine knowledge from both sectors (see Chapters 5 and 7). Furthermore, these established research projects then contribute to overall knowledge enhancement or even disruptions, such as new inventions or new techniques (see Chapter 2). Importantly, such collaborations have an informal element: interest and discussions which occur between researchers and industry can enhance motivation, spark new ideas, and reassure researchers that their research is relevant to practice.

Other examples involve links between business event organisers and local suppliers (Chapters 3 and 6), and collaboration of government agencies and associations with business event organisers (Chapter 4). To illustrate, Chapter 6 explored how a convention centre collaborated with local NGOs to form environmentally sustainable practices in a 'backward' food supply chain.

On the other hand, there were chapters which highlighted that a lack of effective communication and coordination of tasks among two parties could lead to negative outcomes. For instance, Chapter 7 demonstrates that priorities need to be clearly communicated prior to the business event to avoid negative outcomes.

Going further, Figure 10.5b depicts collaboration among multiple parties. A good example is provided in Chapter 3 where conference organisers initiated a corporate social responsibility (CSR) project, The Borneo Orangutan Project, to raise money through corporate donations for the conservation of orangutans in the destination. This initiative was created by an inter-organisational collaboration of an industry association, a government body, a convention bureau, and a local university. The cooperation not only facilitated good relationships and communication, but it also established shared goals for the business event, such as showcasing the city of Kuching in Borneo, Malaysia, and its capability to drive environmental change.

Similarly, Chapters 3, 8 and 9 demonstrate that achieving the objectives and goals of a business event is dependent on effective collaboration among a range of stakeholders, such as convention centres, convention bureaux, industry associations, and government bodies.

## *Stakeholder collaboration managed by organisers*

The case studies demonstrate that effective collaboration among stakeholders leads to the achievement of desired legacies from the business event. Traditional ways to achieve this were to provide sufficient space for conference attendees to connect and engage in discussions, such as coffee breaks (mentioned in Chapter 7), social events (mentioned in Chapter 3) and formalised meetings (Chapters 4, 8 and 9). While individual stakeholders may have planned to leverage specific outcomes, under this approach they are largely incidental.

**Figure 10.6:** Stakeholders' collaboration managed by organisers

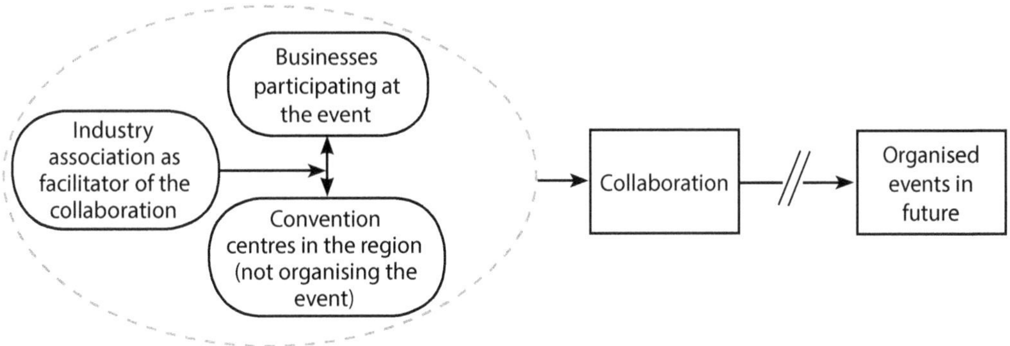


An alternative approach involves organisers directly orchestrating legacy by facilitating various types of connections (Figure 10.6). In particular, Chapter 3 presents an emerging collaboration between participating businesses from the business event industry and three Malaysian convention centres as potential organisers of future business events. A key factor in this process was the industry association, as one of the organising parties, managing the process. Moreover, the collaboration between government bodies (ministries and agencies) and participating event businesses through meetings before and during the conference were leveraged to realise event objectives.

10

The pathways discussed above provide evidence that business events are tapping into what Zahra and George (2002) refer to as recipients' predisposition and readiness to gain new knowledge, leading to positive impacts on the efficacy of knowledge transference. Conferences are optimal sites for enabling such acquisition and facilitating spirals of knowledge (see Chapters 2, 5 and 8) by providing an environment for individuals to collect, acquire or distribute knowledge. They are supporting a constant movement of knowledge interactions which trigger the creation of new spirals. At a micro level, the constant conversions of knowledge between attendees and their interactions during and after the conference amplify the scope for spirals of knowledge. As Chapter 5 points out, conferences deliver the conditions to prompt the different processes of knowledge conversion, informal (rich in tacit knowledge, i.e. socialisation and internalisation), and structural (dense in explicit knowledge, i.e. externalisation and combination). Conferences provide structural processes of knowledge creation through codified and formal language, which is transferred in a systematic way through presentations and keynotes. They also provide a social system in which there are strategic opportunities for exposure, experience, expertise and embedding strategies (Chapter 5), thereby improving knowledge dissemination and utilisation.

## Conclusion

This text has presented a set of case studies demonstrating how conferences, trade shows, exhibitions, professional conference organisers, bureaux and venues from around the globe are making outstanding legacy contributions to economies, communities and the environment – many of which are aligned with the United Nations Sustainable Development Goals. A consideration arises however: how much *more* powerful might such benefits be if event designers were *intentional* about ensuring such legacies are created, managed and monitored? Finally, would more support be provided for business events by governments and business if the industry were able to develop tools, processes and practices which encouraged, supported, measured and leveraged the legacies created?

While many important questions have been discussed, many still remain, and new ones will continue to arise. Future research might explore, for example, the relationships between various legacies or how networking leverages knowledge creation as opposed to other forms of activities during or after the business event. Such interaction of various activities and benefits could be mapped in greater detail via qualitative exploration and then measured quantitatively. Moreover, links and collaborations among various stakeholders should be explored. This would include, for example, links formed between academia and industry, as investigated in Chapters 5 and 7, and the role conference organisers play in facilitating these forms of collaboration. In this way, it would be beneficial to see what specific benefits arise from such collaborations and how they would be managed long term.

As a second area, future research might focus more on the differentiation among various stakeholders. This was presented in Chapters 2 and 7 through sectoral or country of origin differences, distinction between early and late career researchers, or demographic diversity. This type of data could be used for detailed exploration of how specific benefits are perceived by delegates, managed by organisers, and transformed into long-term benefits. It could be achieved by recording various ranges of demographic differences and forming diverse sub-groups, such as early researchers from developing countries. By constructing multiple models of management of business event benefits, organisers could utilise those models to enhance event participation and satisfaction of delegates.

Chapters 2, 3 and 6 referred to the important role of business event support activities, such as engagement of convention centres and delegates in CSR, or providing quality food menus and supporting local suppliers. These activities form unique experiences for conference attendees, and their impact on legacies for various stakeholders should be investigated in multiple contexts. While the impact of these support activities on conference success, discussed in Chapters 2 and 6, is still to be examined, it is clear that all these activities could potentially bring social and environmental benefits to wider communities.

As suggested in the text (Chapters 2, 5 and 7), longitudinal data collection is needed. One example of a best-case scenario would be to record delegates' expectations before the business event, benefits gained during and immediately after the event, and the benefits realised in the long-term. These could be contrasted with the objectives set by conference organisers and their actual strategies to manage benefits from the business event. For instance, how to navigate the benefits, so they lead to the development of a professional career after the conference, remains to be examined. Another relevant question is how greater collaboration between industry and academia can be facilitated and converted into positive long-term outcomes (e.g. funded research project or application of specific technology in practice).

Our final point is for further methodological avenues for future research. For instance, as proposed in Chapter 7, the outliers identified from both qualitative and quantitative data collection should be investigated in greater detail. Moreover, additional comparative studies focusing on several business events and utilising the same research tool are essential (as presented in Chapters 2 and 7). These would help to navigate specific strategies to leverage the benefits of large versus small business events, business events in regional versus city areas, and business events organised by diverse parties.

Business events have the potential to become important instruments in the quest to deliver innovative solutions to current and emerging global challenges. Those responsible for delivering such events have the power to decide just how much of that potential will be realised. The Seven Steps Model was suggested as a guide to maximise this potential. The model can be used by event professionals and the business event industry to ensure their business events are *intentionally* designed to generate legacies, and leverage value for their stakeholders.

## References

Achakulvisut, T., Ruangrong, T., Bilgin, I., Van Den Bossche, S., Wyble, B., Goodman, D. and Kording, K. (2020) Point of view: Improving on legacy conferences by moving online, *Elife*, **9**, e57892.

Berridge, G. (2006) *Events Design and Experience*, London: Routledge. https://doi.org/10.4324/9780080468112

Brown, T. (2008) Design thinking, *Harvard Business Review*, June.

Buchanan, R. (1992) Wicked problems in design thinking, *Design Issues*, **8** (2), 5-21.

Crowther, P. (2014). Investigating design for disassembly through creative practice. In *Intersections International Symposium*, 2014-06-30.

Crowther, P., Orefice, C. and Beard, C. (2018) At work and play: Business events as entrepreneurial spaces, *The International Journal of Entrepreneurship and Innov*ation, **19** (2), 90-99.

Dam, R. and Siang, T. (2020) What is design thinking and why is it so popular? https://www.interaction-design.org/literature/article/what-is-design-thinking-and-why-is-it-so-popular

Edwards, D., Foley, C. and Malone, C. (2017) *The Power of Conferences: Stories of Serendipity, Innovation and Driving Social Change*, Sydney: UTS ePress. doi.org/10.5130/978-0-6481242-0-7

Foley, C., Edwards, D., Jasovska, P. and Hergesell, A (2019) *Business event legacies*, report prepared for Joint Meetings Industry Council, University of Technology Sydney. https://opus.lib.uts.edu.au/bitstream/10453/135365/2/Business%20Event%20Legacies.pdf

Foley, C., Edwards, D. and Schlenker, K. 2014 Business events and friendship: Leveraging the sociable legacies, *Event Management*, **18** (1), 53-64.

Foley, C., Schlenker, K., Edwards, D. and Lewis-Smith, L. (2013) Determining business event legacies beyond the tourism spend: An Australian case study approach, *Event Management: An International Journal*, **17**, 311-322.

Harvard Business Review, Brown, T., Christensen, C., Nooyi, I. and Govindarajan, V. (2020) *HBR's 10 Must Reads on Design Thinking*, Harvard Business Review Press, Tennessee: La Vergne.

Liedtka, J. (2018) Why design thinking works, *Harvard Business Review*.

Stevens, E. (2020) What is design thinking? A comprehensive beginner's guide. https://careerfoundry.com/en/blog/ux-design/what-is-design-thinking-everything-you-need-to-know-to-get-started/

The Iceberg (2019) Legacies of business events. https://www.the-iceberg.org/

Viglione, G. (2020) A year without conferences? How the coronavirus pandemic could change research, *Nature*, **579,** 327-328. https://doi.org/10.1038/d41586-020-00786-y

Zahra, S. A. and George, G. (2002). Absorptive capacity: A review, reconceptualization, and extension. *Academy of Management Review*, **27**(2), 185-203.

# Index


4E framework of training methods 120–121

absorptive capacity, for knowledge 118–120
academic credibility cycle 187
academic scholars 165–196
awareness-raising 34, 54

Ba, and knowledge creation 123
Borneo Orangutan Project 54
business events
  as actors in social systems 214
  research 17
Business Events Value Triangle 92

case studies
  benefits and challenges of hosting conferences for academics and universities 167–185
  China-LAC Business Summit 222–234
  conferences, Sydney 17–33
  EuroHeartCare 2017 105–118
  ICC Sydney: Feeding Your Performance 133–152
  ICCA Congress, Malaysia 47–56
  London Tech Week 77–89
  Swiss Fintech Start-up Corner 199–213
China, investments in Latin America and the Caribbean 222–234
China-LAC Business Summit 222–234
  benefits 229–233
  media impact 231–232
collaboration 26, 27, 33, 59, 68–69
  as legacy 22
community benefits 34–35
community engagement 52–54
conference process 172
conferences
  academic 165–196
  academic, capacity for leveraging benefits 189
  academic, impacts 170–171
  academic, origins 165
  as economic catalysts 16–45
  barriers and challenges for hosts 179–183
  benefits and barriers 173
  benefits for the host researcher 175–177
  benefits for the host university 177–178
  bidding process 181
  delegate perceptions of benefits 113–114
  important for credibility 187
  knowledge acquisition 119
  knowledge creation 122
  knowledge transmission 117
  motivations to attend 111–112
  potential benefits of hosting for universities 191
  Sydney 17–33
Corporate Social Responsibility (CSR) 53, 152–153
  sustainable food procurement 155–156
COVID-19 242
creativitive environments 214

Danish meetings industry 168
delegate motivations 111–112
delegate perceptions of benefits 113–114

design thinkers, attributes 248–249
design thinking 248–251
user-focus 249–250
wicked problems 251–252
developing nations 47–74
dimensions of legacy
application 37
attitudinal 39
knowledge economy 39
ransformative social network 40
diplomacy 221–241

economic benefits
for governments 89
for professional associations 90
for the community 91
short and long term 32
ways of measuring 93–99
economic development
support for local producers 145–148
supporting local small business 146–147
economic diplomacy, China's approach 237
economic impact calculations 80–83
economic impact framework 93
economic impacts, short and long term 85
economic legacies 75–103
economic sustainability in agricultural industry 135
environmental sustainability 134
epistemic community 39
EuroHeartCare 2017 105–118
events, measuring the contribution 4
expenditure zone in benefits calculations 98

Feeding Your Performance 133–152
philosophy and practice 138–140
fintech sector 201
first-tier destinations, defined 63
food miles 142–143
food supply chain 135
keeping short 153–154
social contributions 148–149
funding, access to 34

Geneva, as a fintech hub 201
global buzz
information-rich ecology of conferences 33
International Congress and Convention Centre Association (ICCA) 48
ICCA Congress, Malaysia 47–57
Facebook Live videos 55
ICC Sydney 133–152
innovation 24, 197–220
and knowledge sharing 32
as legacy 21
definition 197
networking at events 213
inscope expenditure 97
categories 98–100
international research conferences 169
International Year of Sustainable Tourism for Development 134

knowledge acquisition
experience strategies 120
maximising 38
knowledge assimilation
expertise strategies 121
knowledge-based events
impact on host area 215
knowledge conversion modes 124
knowledge creation, through conferences 122
knowledge dissemination,
exposure strategies 120
knowledge economy 29–30, 39, 169–170
knowledge gain 109
knowledge sharing
absorptive capacity 118
and innovation 32
and organisational development 104–131
defined 109
tacit and codified 36
knowledge, tacit and explicit 124
knowledge-transfer processes
Nonaka and Takeuchi model 37

knowledge transmission from conferences 117–118
knowledge use, embedding strategies 121
Kuching, as second-tier destination 49

Latin American and Caribbean (LAC) countries 223
legacies
  challenges in leveraging 246–249
  development of benefits over time and space 257
  lapse of time before realisation 29
  leveraging 254–259
  long-term indirect benefits 255–256
  reinforcement of benefits 254–255
  stakeholder collaboration 257
legacy creation, designing 247–252
legacy outcomes
  and stakeholders 244
  innovative event design 244
  planning for 243–262
  pre-planned 244
legacy planning 95
  seven-step framework 252–255
local food 135–138
  supporting producers 145–148
local-global links 245
London Tech Week 77–89
longer-term impacts 30, 31, 61–62
'long tail' outcomes 33

Malaysia 47–57
  government engagement 51
  gross national income breakdown 49

networking 33, 110
  and innovation 213
  local-global, shaped by events 214
new money 97
Nonaka and Takeuchi
  knowledge-transfer processes 37
  theory of organisational knowledge creation 123–125

Olympic Games, economic legacies 94
ommunity of practice 39

partnerships
  and sustainable development goals 66–68
personal recognition 111
play, role for entrepreneurial outcomes 213
playfulness in an event context 213
professional development 51, 110
Punta del Este 223
  Convention & Exhibition Center 225–226

reputation benefits 86
research, value of business events 17
return on investment (ROI), measurement 227

second-tier cities 46
  advocacy 56
  defined 63
sector development, as legacy 22–23
sentiment analysis 83
showcasing 34
showcasing expertise 27
SIBOS (Swift International Banking Operations Seminar) 199
social capital
  communities 26
  individuals 25
social media metrics 204, 209
social network 40
soft power 221–241
  and economic development 237–238
  and national culture 238
  basis of 236
  China's use of 236–238
  defined 235
  implementation 236
spirals of knowledge 122–126
stakeholder benefits 31–35
stakeholder collaboration 257–260
stakeholder roles 61–62

stakeholders 60–61, 253-254
startups 198
  challenges 216
  driving innovation and employment 215
stewardship 143–144
supply chains 132–164
survey techniques 19–20
sustainability 132–164
  in agriculture 141–142
  stewardship 143–144
  waste management 142
sustainable development 46–74
sustainable food procurement
  through CSR 155–156
Swiss Fintech Start-up Corner 199–213
Switzerland as a fintech hub 201
tourism contribution, tip of the iceberg 89
tourism contribution to events 4
tourism degrees 3
tourism/events partnership 3–4
training methods, 4E framework 120
Twitter, measuring impact 228

United Nations Sustainable Development
  Goals 56, 64–66
Uruguay 223

virtuous dynamic of conference legacy 35–36
visitor economy, contributions to 96–99

wicked problems 251–252

www.ingramcontent.com/pod-product-compliance
Ingram Content Group UK Ltd.
Pitfield, Milton Keynes, MK11 3LW, UK
UKHW051847100926
8556IPUK00005B/506

9 781911 635291